AF688235

Table of Contents

1 Introduction

The convergence of seemingly disparate fields often sparks unexpected innovation. Consider the intersection of artificial life and bioluminescence, two areas brimming with potential. Artificial life, the quest to create life-like systems in silico or in vitro, offers a unique lens through which to study and understand the complex processes of life. Bioluminescence, the production and emission of light by living organisms, provides a captivating display of nature's ingenuity. The spectral overlap between artificially generated light and naturally occurring bioluminescence opens exciting possibilities for research and application. This book delves into this fascinating interplay, exploring the potential for synchronicity between these two domains.

The genesis of artificial life research stems from a desire to understand the fundamental principles of life itself. By attempting to recreate life's essential characteristics in artificial systems, we can gain insights into the very nature of living organisms. Early pioneers in the field envisioned self-replicating machines and evolving algorithms, paving the way for the development of complex simulations and synthetic biological constructs. This pursuit has led to advancements in fields like robotics, computer science, and synthetic biology, providing valuable tools for studying complex systems and designing innovative solutions.

Bioluminescence, on the other hand, has captivated scientists and nature enthusiasts alike for centuries. From the mesmerizing glow of fireflies to the ethereal shimmer of deep-sea creatures, bioluminescence showcases the remarkable diversity of life on Earth. The underlying biochemical mechanisms responsible for this natural light production vary significantly across different organisms, each adapted to specific ecological niches and purposes. Understanding these mechanisms provides a rich source of inspiration for developing novel light-emitting technologies and exploring the intricate interplay between light and life.

The spectral characteristics of light emitted by both artificial organisms and bioluminescent organisms hold a key to unlocking their potential for interaction. The electromagnetic spectrum encompasses a vast range of wavelengths, from radio waves to gamma rays. Visible light, the portion of the spectrum we can perceive with our eyes, represents a small but crucial window into the world. The specific wavelengths emitted by a light source determine its color and can influence its interaction with other organisms and the environment. The overlap between the emission spectra of artificial and natural light sources creates opportunities for communication, control, and manipulation of biological systems.

Imagine a scenario where artificial organisms, equipped with precisely tuned light-emitting components, can communicate with bioluminescent bacteria. By mimicking the spectral patterns of bacterial bioluminescence, these artificial organisms could potentially influence bacterial behavior, opening up avenues for targeted drug delivery or environmental remediation. Similarly, bioluminescent organisms could be used as living sensors, their light output modulated by environmental factors like pollutants or temperature changes. Artificial sensors could then detect and interpret these changes, providing real-time monitoring of environmental conditions.

The potential applications of spectral synchronicity between artificial organisms and bioluminescence extend beyond the laboratory. In the realm of bio-inspired robotics, bioluminescent systems can serve as a model for developing novel light-based communication and navigation strategies for underwater vehicles. The ability to generate and detect specific light signals in the dark depths of the ocean could revolutionize underwater exploration and research. Moreover, understanding the intricate mechanisms of bioluminescence can inspire the design of new materials and devices with enhanced light-emitting properties, leading to advancements in lighting technology and displays.

The field of spectral synchronicity between artificial organisms and bioluminescence is still in its nascent stages, but the potential for groundbreaking discoveries and applications is immense. As we delve deeper into the intricacies of light production and interaction in both artificial and biological systems, we unlock new possibilities for understanding and manipulating the world around us. The journey into this unexplored territory promises to be a fascinating and illuminating one.

1.1 Genesis of Artificial Life

The pursuit of creating artificial life began not in a sterile laboratory, but in the human imagination. Long before the tools of biotechnology existed, philosophers and storytellers pondered the nature of life itself, questioning whether it could be replicated or even improved upon. Myths and legends are replete with artificial beings, from golems of clay to automatons of clockwork, reflecting a deep-seated curiosity about the boundary between inanimate matter and living organisms. This fascination persisted through the scientific revolution, eventually finding expression in the burgeoning field of computer science. Early pioneers recognized the potential of computers not just as calculating machines, but as platforms for simulating complex systems, including the very processes of life itself.

The quest to understand and recreate life artificially gained momentum with the advent of cybernetics in the mid-20th century. Researchers began exploring the parallels between biological systems and computational processes, drawing inspiration from the self-regulating mechanisms found in living organisms. This period saw the development of the first self-replicating computer programs, simple algorithms capable of copying themselves, hinting at the possibility of digital "life." These early experiments laid the groundwork for later, more sophisticated simulations of evolutionary processes and artificial ecosystems.

The field took a decisive step forward with the formalization of "artificial life" as a distinct research area in the 1980s. Scientists from diverse disciplines – computer science, biology, physics, and even philosophy – convened to explore the potential of creating life in silico, within the digital realm of computers. This marked a shift from simply mimicking biological processes to actually generating new forms of life, albeit in a virtual environment. Early artificial life experiments focused on creating simple, self-replicating digital organisms that could evolve and adapt to changing conditions within their simulated worlds.

The initial artificial life forms were often abstract, represented by strings of code or simple geometrical shapes interacting within a defined digital space. However, the underlying principles they embodied were deeply rooted in biological concepts, such as self-replication, mutation, and natural selection. By observing the behavior of these digital organisms, researchers hoped to gain insights into the fundamental processes that govern the evolution of life itself, both on Earth and potentially elsewhere in the universe. The challenge lay in designing systems complex enough to exhibit emergent behavior – unpredictable and novel properties arising from the interactions of simpler components – yet simple enough to be computationally tractable.

The development of more powerful computers and sophisticated algorithms allowed researchers to create increasingly complex artificial life simulations. These virtual worlds became populated with digital organisms exhibiting a wide range of behaviors, from simple foraging strategies to complex social interactions. The emergence of cooperation and competition within these digital ecosystems mirrored the dynamics observed in natural biological systems, providing a powerful tool for studying the evolution of altruism, predator-prey relationships, and other fundamental ecological principles.

As the field matured, researchers began to explore the possibility of creating artificial life not just in software, but also in hardware, using robotics and synthetic biology. This ambition led to the development of physical robots capable of exhibiting lifelike behaviors, such as navigation, obstacle avoidance, and even simple forms of learning. Simultaneously, advances in synthetic biology opened up new avenues for creating artificial cells and other biological building blocks from scratch, blurring the lines between the living and the engineered. The focus shifted from purely computational simulations to the creation of tangible, physical instantiations of artificial life.

The development of these more tangible forms of artificial life raised profound philosophical and ethical questions. What does it mean to create life? What responsibilities do we have towards artificial organisms that exhibit intelligence or even consciousness? These questions continue to be debated within the field, driving researchers to consider the broader implications of their work. The ethical considerations surrounding the creation of artificial life have become increasingly important as the field progresses, prompting discussions about the potential risks and benefits of such technologies. As artificial organisms become more sophisticated, the need for careful ethical guidelines and regulations becomes ever more critical.

The genesis of artificial life represents a convergence of ancient curiosity and cutting-edge scientific inquiry. From philosophical musings on the nature of life to the creation of complex simulations and physical robots, the field has traversed a remarkable journey. As we continue to push the boundaries of what is possible, the quest to understand and create artificial life promises to yield profound insights into the very essence of life itself, while simultaneously presenting us with unprecedented ethical and societal challenges. The ongoing exploration of artificial life is not just a scientific endeavor, but a deeply human one, reflecting our enduring fascination with the origins and possibilities of life in all its forms.

1.2 Bioluminescence Overview

Bioluminescence, the production and emission of light by living organisms, is a captivating phenomenon observed across a diverse range of species. From the shimmering glow of fireflies to the ethereal luminescence of deep-sea creatures, this remarkable ability has evolved multiple times, serving various biological functions. Understanding the underlying mechanisms and spectral characteristics of bioluminescence is crucial for appreciating its role in nature and for exploring its potential applications in artificial systems. This chapter provides a comprehensive overview of bioluminescence, encompassing its diverse forms, underlying chemical processes, and ecological significance.

The vast majority of bioluminescent organisms are found in marine environments, particularly in the deep ocean where sunlight cannot penetrate. Here, bioluminescence plays a crucial role in communication, predation, and defense. For example, anglerfish use bioluminescent lures to attract prey, while some deep-sea shrimp expel clouds of luminous material to confuse predators. In shallower waters, dinoflagellates, microscopic algae, create stunning bioluminescent displays when disturbed, often triggered by the movement of waves or boats. Terrestrial bioluminescence, though less common, is exemplified by fireflies, which use precisely timed flashes for attracting mates. Certain fungi, like the ghostly glowing mushrooms found in decaying wood, also exhibit bioluminescence, although the ecological function in these cases is less well understood. This diversity of habitats and purposes underscores the remarkable adaptability of bioluminescence.

At the heart of bioluminescence lies a chemical reaction involving a light-emitting molecule called luciferin and an enzyme called luciferase. Luciferases catalyze the oxidation of luciferin, a process that releases energy in the form of light. The specific luciferin-luciferase system employed varies across different organisms, resulting in a wide range of emitted colors, from blue and green to yellow and red. This variation in color is determined by the chemical structure of the luciferin and the specific configuration of the luciferase enzyme. Some organisms, like certain jellyfish, utilize a different mechanism involving photoproteins. Photoproteins are proteins that contain a pre-bound luciferin and require only a trigger, often calcium ions, to initiate the light-emitting reaction. The unique characteristics of each bioluminescent system contribute to the incredible diversity of light displays observed in nature.

Examining the chemical pathways underlying bioluminescence reveals a fascinating interplay of enzyme kinetics and energy transfer. In the luciferin-luciferase reaction, the luciferase enzyme binds to both luciferin and oxygen, bringing them into close proximity to facilitate the oxidation reaction. The energy released during this reaction excites the luciferin molecule to a higher energy state. As the excited luciferin returns to its ground state, it releases the excess energy as a photon of light. The precise wavelength of the emitted light, and thus its color, is determined by the energy difference between the excited and ground states of the luciferin molecule. The efficiency of this process is remarkably high, with very little energy lost as heat, making bioluminescence a highly efficient form of light production.

The ecological significance of bioluminescence extends beyond simple signaling. In some marine ecosystems, bioluminescence plays a vital role in symbiotic relationships. For instance, certain deep-sea fish harbor bioluminescent bacteria in specialized light organs. The fish provide the bacteria with nutrients and a protected environment, while the bacteria provide the fish with a source of light for attracting prey or mates. These symbiotic partnerships showcase the complex interplay between different organisms and the multifaceted roles of bioluminescence in shaping ecological communities.

Furthermore, bioluminescence serves as a valuable tool for scientific research. Researchers use bioluminescent molecules as markers to study gene expression, track cellular processes, and detect specific biomolecules. The sensitivity and specificity of bioluminescent assays make them powerful tools in various fields, from biomedical research to environmental monitoring. By harnessing the power of bioluminescence, scientists can gain valuable insights into the intricate workings of biological systems.

In closing, bioluminescence represents a remarkable example of biological adaptation and chemical ingenuity. The diverse mechanisms, spectral characteristics, and ecological functions of bioluminescence highlight the fascinating interplay between chemistry, biology, and the environment. From the depths of the ocean to the forests of the night, bioluminescence illuminates the wonders of the natural world and provides a rich source of inspiration for scientific exploration. Understanding the fundamental principles of bioluminescence is essential for the later chapters of this book, which explore the interplay between natural bioluminescence and its artificial counterparts. This foundation will enable us to delve into the intricate world of spectral synchronicity and its potential applications.

1.3 Spectral Overlap

The concept of spectral overlap, as it pertains to artificial organisms and bioluminescence, lies at the heart of this exploration. It signifies the degree to which the emission spectra of artificially generated light, produced by our engineered constructs, coincide with the bioluminescent spectra found in nature. This overlap has profound implications for potential interactions and communication between artificial and biological systems, forming the basis for achieving spectral synchronicity.

Imagine a scenario where robotic probes, equipped with bioluminescent mimics, seamlessly integrate into a marine environment, communicating with and studying organisms without disrupting their natural behaviors. This vision relies heavily on precise spectral matching, allowing for covert observation and interaction within delicate ecosystems.

Understanding spectral overlap requires delving into the intricacies of light emission. Each bioluminescent organism, from the firefly to the deep-sea anglerfish, produces light within a specific range of wavelengths, creating a unique spectral signature. These signatures are not merely arbitrary; they are products of evolutionary pressures, finely tuned for specific functions like attracting mates, luring prey, or deterring predators. Artificial organisms, if they are to interact effectively with these biological counterparts, must be capable of generating light within the same or closely overlapping spectral bands. Consider the firefly, whose flashing bioluminescence serves as a complex mating signal. An artificial organism aiming to study firefly communication would need to replicate not just the timing of the flashes, but also the precise spectral composition of the emitted light.

Achieving this spectral matching presents a significant engineering challenge. While we can generate light across a broad spectrum using various technologies, replicating the precise spectral nuances of biological systems is complex. The specific chemical reactions that produce bioluminescence, involving luciferins, luciferases, and other photoproteins, are incredibly sophisticated and difficult to reproduce artificially. Moreover, the environment plays a crucial role in shaping the emitted spectrum. Factors such as temperature, pH, and the presence of certain ions can subtly alter the wavelength of emitted light, adding another layer of complexity to the challenge of spectral mimicry. Researchers are actively exploring different approaches, including bio-inspired materials and synthetic biology, to overcome these hurdles and create artificial light sources with the necessary precision and control.

Beyond simply mimicking existing bioluminescent spectra, the concept of spectral overlap opens doors to entirely new possibilities. We can envision artificial organisms emitting light in spectral regions not found in nature, potentially opening channels of communication with biological systems that are sensitive to those wavelengths. This could allow us to explore fundamental biological processes in new ways, probing cellular responses to light signals that have never been encountered before.

However, this approach requires careful consideration of potential unintended consequences. Introducing novel light signals into an ecosystem could have unforeseen impacts, disrupting established communication networks or triggering unexpected biological responses.

The practical applications of spectral overlap extend beyond scientific inquiry. Imagine swarms of bioluminescent robots, emitting light synchronized with the surrounding marine organisms, used for underwater mapping and exploration. These robots could navigate complex underwater environments, communicating with each other and collecting data without disturbing the delicate balance of the ecosystem. In another scenario, bioluminescent sensors, calibrated to match the spectral signatures of specific pollutants, could provide real-time monitoring of water quality, offering early warnings of environmental contamination.

Furthermore, the study of spectral overlap provides a unique lens through which to investigate the fundamental principles of biological communication. By analyzing the intricate interplay of light signals between organisms, we can gain deeper insights into the evolutionary pressures that shaped these communication systems. This knowledge can, in turn, inform the design of more sophisticated artificial organisms, capable of engaging in complex interactions with their biological counterparts. As we delve deeper into the realm of spectral overlap, we uncover not just the potential for technological innovation, but also a deeper understanding of the intricate web of life that surrounds us. The ability to bridge the gap between the artificial and the biological, through the precise manipulation of light, promises to unlock a new era of scientific discovery and technological advancement. This intricate dance of light and life, guided by the principles of spectral overlap, holds the key to unlocking a deeper understanding of the natural world and our place within it. By carefully studying and mimicking the spectral signatures of bioluminescent organisms, we can create artificial systems that seamlessly integrate into the biological world, opening up exciting new possibilities for scientific exploration, technological innovation, and a deeper appreciation for the wonders of bioluminescence.

2 Artificial Organisms

Artificial life, distinct from naturally occurring life, encompasses artificially created systems that exhibit lifelike characteristics. These systems, often realized through computer simulations or robotic constructions, provide a platform for exploring the fundamental principles of life, including self-organization, adaptation, and evolution. One particular area of focus lies in recreating bioluminescence – the emission of light by living organisms – within these artificial constructs. This endeavor presents unique opportunities for understanding and potentially harnessing the complex processes underlying this natural phenomenon. By simulating bioluminescence, researchers can investigate the interplay of chemical reactions, genetic regulation, and environmental factors that contribute to light production in living organisms.

Simulating bioluminescence in artificial organisms involves a multifaceted approach, requiring expertise in diverse fields such as computer science, engineering, and biology. One common method involves computational modeling of the biochemical pathways responsible for light production. These models, based on known or hypothesized reactions, can simulate the dynamics of enzyme-substrate interactions and the resulting light emission. Parameters like enzyme concentrations, reaction rates, and environmental conditions can be adjusted within the model to explore their impact on bioluminescent output. Such simulations can provide insights into the regulatory mechanisms that control light production and help identify key factors influencing its intensity and duration.

Another approach leverages synthetic biology to engineer artificial cells or organisms capable of expressing bioluminescent proteins. This involves incorporating genes encoding luciferases – the enzymes that catalyze light-emitting reactions – into the genetic material of the artificial organism. The expression of these genes leads to the production of luciferases, which, in the presence of their respective substrates (luciferins), generate light. The choice of luciferase-luciferin system can influence the color and intensity of the emitted light, allowing for customization of the artificial bioluminescence.

The engineering challenges extend beyond genetic manipulation. Creating artificial organisms capable of sustained and controlled bioluminescence necessitates careful consideration of energy sources, substrate delivery, and waste removal. Researchers are exploring various strategies to address these challenges, including the development of microfluidic devices that can mimic the circulatory systems of living organisms, providing a continuous supply of nutrients and removing metabolic byproducts. These microfluidic platforms also offer precise control over environmental factors, enabling the study of how temperature, pH, and other parameters influence bioluminescence.

Furthermore, the physical design of the artificial organism plays a crucial role in shaping the emitted light. The arrangement of light-emitting components, the presence of reflective or refractive structures, and the overall morphology of the organism can influence the directionality, intensity, and spectral properties of the emitted light. Researchers are drawing inspiration from naturally occurring bioluminescent organisms, such as fireflies and deep-sea creatures, to design artificial organisms with optimized light emission patterns. For example, mimicking the intricate lantern structures of fireflies could lead to the development of highly efficient artificial light sources.

The successful integration of bioluminescence into artificial organisms has far-reaching implications for various fields. In robotics, bioluminescent artificial organisms could serve as autonomous light sources for underwater exploration, eliminating the need for external illumination. They could also be used in environmental monitoring, acting as biosensors that respond to changes in pollution levels or other environmental parameters by altering their light emission. The development of bioluminescent artificial organisms represents a significant step toward creating truly lifelike artificial systems, blurring the lines between the natural and the artificial. As we continue to refine our ability to simulate and control bioluminescence in artificial life, we open up exciting possibilities for both scientific discovery and technological innovation. The future of this field promises to be bright, both literally and figuratively.

The spectral characteristics of the artificially generated light are carefully analyzed and compared with the emission spectra of naturally bioluminescent organisms. This analysis enables researchers to understand the similarities and differences between artificial and natural bioluminescence, providing insights into the underlying mechanisms and evolutionary pressures that shape light production in living organisms. By studying the spectral overlap between artificial and natural light emission, scientists can explore the potential for communication or interaction between artificial organisms and their biological counterparts.

Building these simulations involves considering multiple factors. The specific type of luciferin-luciferase reaction being modeled, for example, dictates the chemical kinetics and resulting spectral properties. The model must account for the efficiency of the reaction, the concentration of reactants, and any quenching mechanisms that might reduce light output. Furthermore, the physical environment of the simulation plays a crucial role. Factors such as temperature, pH, and the presence of other molecules can significantly impact the reaction rate and light emission. Accurately simulating these conditions is essential for creating a realistic model of bioluminescence.

Beyond simple biochemical reactions, some models incorporate the spatial organization of light-emitting structures. In many bioluminescent organisms, specialized cells or organs house the light-producing machinery. These structures often incorporate reflectors or lenses that shape and direct the emitted light. Simulating these complex structures requires sophisticated computational techniques, such as ray tracing or finite element analysis, to accurately capture the interaction of light with the surrounding tissues. These advanced models provide a more complete picture of bioluminescence, revealing how structural adaptations influence light emission patterns and signal propagation.

2.1 Defining Artificial Life

Artificial life, a field brimming with both scientific rigor and philosophical intrigue, seeks to understand the fundamental principles of life by creating it anew, in silico or in vitro. This pursuit isn't about replicating existing biological organisms, but rather about abstracting the core principles of life – organization, metabolism, adaptation, reproduction – and implementing them in novel substrates. We strive to understand the essence of what makes something "alive" by building it from the ground up. Think of it as exploring the blueprint of life, not just copying an existing structure.

A key distinction must be drawn between "life-as-we-know-it," the carbon-based life that flourishes on Earth, and "life-as-it-could-be." Our planet's biology provides a rich, but ultimately singular example of life. Artificial life expands this perspective by exploring the potential for life to exist in forms radically different from what we observe in nature. This could include digital organisms evolving within a computer simulation, or synthetic biological constructs built from non-organic materials. By breaking free from the constraints of terrestrial biology, artificial life offers a glimpse into the vast, unexplored landscape of possible life forms.

Several approaches to creating artificial life are currently being explored. Software simulations, often called "digital organisms" or "artificial chemistries," allow researchers to model the evolution and behavior of simple life forms in virtual environments. These simulations can test hypotheses about the origins of life and the factors that drive its complexity. Another approach involves creating synthetic biological systems, building new life forms from the molecular level up, using both organic and synthetic components. These efforts blur the lines between biology and engineering, pushing the boundaries of what we consider "living."

The study of artificial life isn't limited to abstract models and laboratory experiments. It also draws inspiration from robotics, where researchers attempt to imbue machines with life-like qualities. These "robots" may not be "alive" in the traditional sense, but they can exhibit complex behaviors, adapt to their environment, and even interact with other "organisms" in their artificial world. This interplay between artificial life and robotics offers valuable insights into the relationship between form, function, and behavior.

Defining artificial life raises profound philosophical questions about the very nature of life itself. What are the minimal requirements for something to be considered alive? Can consciousness arise in non-biological systems? These questions, while deeply challenging, are at the heart of the artificial life enterprise. By grappling with these questions, we gain a deeper appreciation for the complexity and wonder of life, both as we know it and as it could be. The search for artificial life is not just a scientific endeavor, but also a philosophical journey that compels us to re-evaluate our place in the cosmos.

Building artificial life demands a multidisciplinary approach, drawing on expertise from computer science, biology, chemistry, physics, and even philosophy. Computer scientists develop the algorithms and simulations that drive artificial life experiments. Biologists contribute their knowledge of biological processes, providing inspiration and guidance for designing artificial organisms. Chemists and physicists explore the physical and chemical properties of the materials used to build synthetic life. Philosophers grapple with the ethical and conceptual implications of creating artificial life. This convergence of disciplines creates a vibrant intellectual ecosystem, fostering innovation and pushing the boundaries of scientific inquiry.

Consider the intriguing possibility of artificial organisms inhabiting virtual worlds, evolving and interacting according to their own internal logic. Imagine synthetic biological systems exhibiting emergent properties, unexpected behaviors arising from the interactions of their constituent parts. Envision robots that learn, adapt, and even reproduce, blurring the lines between machine and organism. These are not just futuristic fantasies, but tangible goals within the realm of artificial life research.

The implications of creating artificial life are far-reaching and profound. By understanding the fundamental principles of life, we can potentially develop new technologies for medicine, agriculture, and environmental remediation. Artificial life could lead to the creation of self-healing materials, bio-inspired robots for exploration and rescue, and even new forms of energy production. Beyond these practical applications, artificial life offers a unique lens through which to explore the deepest mysteries of existence, challenging our assumptions about life, consciousness, and our place in the universe.

Ultimately, the quest for artificial life is a testament to human curiosity and our relentless pursuit of knowledge. It is a journey of discovery, not just of the world around us, but also of ourselves. By creating artificial life, we gain a deeper understanding of what it means to be alive, pushing the boundaries of science and philosophy in the process. This ongoing exploration continues to reshape our understanding of life's intricate tapestry and challenges us to ponder the very essence of existence. This search opens new doors to scientific discovery and pushes the limits of human ingenuity, fostering innovation and inspiring new generations of scientists and thinkers. The potential rewards are immense, promising to revolutionize our understanding of life and our place in the cosmos.

2.2 Simulating Bioluminescence

Recreating the mesmerizing glow of bioluminescence in artificial organisms presents a unique set of challenges. We must consider not only the chemical processes involved, but also the intricate interplay of light emission, scattering, and absorption within the artificial organism's structure. This chapter will delve into the practicalities of simulating bioluminescence, examining various methods and their respective advantages and limitations. Consider the intricacies of mimicking the flashing patterns of fireflies or the subtle glow of deep-sea creatures – each requiring a different approach.

One common method involves using light-emitting diodes (LEDs) to approximate the spectral characteristics of bioluminescent organisms. LEDs offer precise control over color and intensity, making them suitable for simulating a wide range of bioluminescent phenomena. By carefully selecting LEDs with specific emission wavelengths and modulating their output, we can create dynamic light patterns that resemble the natural world. Furthermore, the compact size and low power consumption of LEDs make them ideal for integration into small-scale artificial organisms. However, LEDs lack the inherent softness and diffusion of natural bioluminescence. Their light tends to be more directional and less organically dispersed.

To address this limitation, we can explore the use of chemiluminescent materials. These materials produce light through chemical reactions, similar to how natural bioluminescence occurs. By encapsulating chemiluminescent compounds within a suitable matrix, we can achieve a more diffuse and naturally distributed glow. The challenge lies in controlling the reaction rate and duration, as well as ensuring the long-term stability of the chemiluminescent material. While offering a closer approximation to the organic quality of natural bioluminescence, chemiluminescence can be less predictable and controllable than LED-based approaches.

Another promising avenue involves bioluminescent proteins, such as green fluorescent protein (GFP) and its variants. These proteins can be genetically engineered into artificial organisms, enabling them to produce their own light. This approach offers the potential for highly realistic and integrated bioluminescence. However, it requires sophisticated genetic engineering techniques and careful consideration of the host organism's physiology. Maintaining the stability and functionality of bioluminescent proteins within an artificial system can also be a significant hurdle.

The choice of simulation method depends largely on the specific application and the desired level of realism. For applications where precise control over light emission is paramount, LEDs may be the preferred option. For scenarios requiring a more natural and diffuse glow, chemiluminescence or bioluminescent proteins may be more suitable. Consider a robotic jellyfish designed for underwater exploration – the gentle pulsating light achieved through chemiluminescence could enhance its camouflage and communication capabilities. Alternatively, an artificial firefly designed for swarm robotics might benefit from the precise control offered by LEDs to synchronize flashing patterns and facilitate coordinated behavior.

Beyond replicating the visual appearance of bioluminescence, we must also consider its spectral characteristics. Natural bioluminescence exhibits a remarkable diversity of emission spectra, ranging from blue-green in marine organisms to yellow and red in terrestrial species. Accurately reproducing these spectra is crucial for simulating the ecological function of bioluminescence, such as communication, camouflage, and prey attraction. Spectral matching requires careful selection of light sources and materials, taking into account the absorption and scattering properties of the surrounding environment.

Furthermore, the temporal dynamics of bioluminescence play a significant role in its ecological function. Some organisms emit continuous light, while others produce flashes or pulsations. These temporal patterns are often species-specific and serve various purposes, such as attracting mates or deterring predators. Simulating these temporal dynamics requires sophisticated control systems that can modulate the intensity and duration of light emission. Imagine an artificial anglerfish luring its prey with a precisely timed bioluminescent flash – the timing and intensity of the flash are critical for its success.

Finally, consider the spatial distribution of light within the artificial organism. In many bioluminescent organisms, light is not uniformly distributed but is concentrated in specific organs or tissues. This localized emission can enhance the effectiveness of bioluminescence for signaling or camouflage. Replicating this spatial distribution requires careful design and engineering of the artificial organism's light-emitting structures. Think of the intricate patterns of light displayed by deep-sea cephalopods – these patterns are crucial for communication and camouflage in their dark environment.

Successfully simulating bioluminescence requires a multidisciplinary approach, combining knowledge of biology, chemistry, physics, and engineering. By carefully considering the visual, spectral, temporal, and spatial characteristics of natural bioluminescence, we can create artificial organisms that not only mimic its aesthetic beauty but also its ecological function. As we continue to refine our understanding of bioluminescence and develop new technologies, the potential for creating truly lifelike artificial organisms becomes increasingly tangible.

3 Bioluminescent Systems

Bioluminescence, the production and emission of light by living organisms, manifests in a dazzling array of systems across the biological world. Understanding these diverse mechanisms is crucial to appreciating the complexity of spectral overlap and synchronicity between natural and artificial light. Luciferin-luciferase reactions form the foundation of many bioluminescent systems. This involves a light-emitting molecule, luciferin, which is oxidized by an enzyme, luciferase, in the presence of oxygen. The specific structure of the luciferin and luciferase molecules dictates the color of the emitted light, ranging from blue-green to red. Subtle variations in these molecular structures, even within closely related species, can lead to distinct spectral signatures, highlighting the evolutionary adaptability of bioluminescence.

Photoproteins offer an alternative mechanism for light production. These proteins, such as aequorin found in certain jellyfish, bind to a luciferin-like molecule and oxygen, forming a stable complex. Upon binding calcium ions, a conformational change triggers the oxidation of the bound molecule, resulting in a flash of light. Unlike luciferin-luciferase systems, which require a continuous supply of luciferin, photoproteins contain all the necessary components for light emission preassembled, enabling rapid and intense bursts of bioluminescence. The control over calcium ion concentration allows organisms to modulate the intensity and timing of these flashes.

Bacterial bioluminescence, predominantly observed in marine species, often involves a more complex system known as the Lux operon. This cluster of genes encodes not only the luciferase enzyme but also proteins responsible for synthesizing and recycling the luciferin molecule. The regulation of these genes is influenced by environmental factors such as nutrient availability and population density, leading to phenomena like quorum sensing, where bacteria collectively regulate gene expression based on their population size. This collective behavior can result in synchronized bioluminescent displays, transforming entire swathes of the ocean into shimmering fields of light. The study of bacterial bioluminescence offers valuable insights into bacterial communication and ecological dynamics.

Fungal bioluminescence, though less common than bacterial or marine examples, presents unique characteristics. Certain fungi, particularly those inhabiting decaying wood, emit a continuous, often greenish light. The exact biological function of fungal bioluminescence remains somewhat elusive, with hypotheses ranging from attracting insects for spore dispersal to a metabolic byproduct with no specific adaptive role. The luciferases in fungi exhibit distinct structural differences from those found in other organisms, suggesting independent evolutionary origins. The investigation of fungal bioluminescence opens up new avenues for understanding the evolution and ecological significance of this phenomenon in terrestrial ecosystems.

Marine bioluminescence showcases the incredible diversity and adaptability of light-emitting systems. From the mesmerizing flashes of dinoflagellates, which illuminate breaking waves, to the complex light organs of deep-sea fish, the ocean teems with bioluminescent life. These organisms utilize light for a variety of purposes, including camouflage, attracting prey, and communication. Deep-sea creatures, in particular, have evolved intricate bioluminescent adaptations to navigate and communicate in the darkness of the abyss. Some species harbor symbiotic bacteria within specialized light organs, while others produce their own luciferins and luciferases. The study of marine bioluminescence provides a window into the complex ecological interactions and evolutionary pressures shaping life in the deep ocean. Furthermore, the diverse spectral characteristics of marine bioluminescence, ranging from blue to red, offer a rich palette for understanding the interplay of light and life in the marine environment. This intricate web of light-based interactions underscores the importance of studying bioluminescence in its ecological context. The examination of these natural systems informs the design and implementation of artificial bioluminescent systems, paving the way for exciting applications in various fields.

3.1 Luciferin-Luciferase Reactions

The magic of bioluminescence, the production of light by living organisms, often boils down to a deceptively simple chemical reaction: the interaction of luciferin and luciferase. This chapter delves into the intricacies of these luciferin-luciferase reactions, exploring the diversity of these systems across various organisms and highlighting their significance in the broader context of bioluminescent phenomena. We will unravel the elegant molecular dance that gives rise to living light.

Luciferin, broadly speaking, is the light-emitting molecule. Its structure varies significantly across different species, leading to the breathtaking array of colors observed in bioluminescence – from the ethereal blue of deep-sea creatures to the vibrant green of fireflies. Think of luciferin as the fuel, storing the potential energy waiting to be released as photons. This release, however, requires a catalyst: luciferase. Luciferases are enzymes, biological catalysts that accelerate the oxidation of luciferin. This oxidation reaction transforms the chemical energy stored within luciferin into light energy. The specific mechanism of this transformation differs depending on the specific luciferin-luciferase pair involved. In fireflies, for example, the reaction involves the oxidation of firefly luciferin by firefly luciferase in the presence of adenosine triphosphate (ATP), the cellular energy currency, and oxygen. This process produces an excited state of the luciferin molecule, which then returns to its ground state by emitting a photon of light, typically in the yellow-green range. The efficiency of this reaction is remarkably high, with nearly all the energy converted into light, minimizing heat production. This "cold light" is a testament to the evolutionary refinement of these bioluminescent systems.

Beyond fireflies, a vast array of organisms utilize different luciferin-luciferase systems. In click beetles, a distinct luciferin molecule interacts with its corresponding luciferase, producing a range of colors from green to orange. Dinoflagellates, microscopic marine organisms, employ a luciferin derived from chlorophyll, contributing to the sparkling bioluminescence often seen in ocean waves. Each system, with its unique luciferin-luciferase pairing, represents a distinct evolutionary solution to the challenge of producing light.

Understanding the specific mechanisms of these reactions is crucial. The structure of the luciferin molecule dictates the color of the emitted light. Subtle variations in the luciferin structure can lead to significant shifts in the emission spectrum. Similarly, the luciferase enzyme plays a critical role in determining the efficiency and kinetics of the reaction. Its structure influences how quickly the reaction occurs and how much light is produced. By studying the interplay between luciferin and luciferase, we can begin to decipher the intricate chemical language of bioluminescence.

Furthermore, the study of luciferin-luciferase reactions is not limited to understanding naturally occurring bioluminescence. These systems have found widespread applications in various scientific fields. Researchers have harnessed the power of firefly luciferase to develop sensitive assays for detecting ATP, a crucial indicator of cellular activity. This technique has applications in drug discovery, environmental monitoring, and food safety testing. Other luciferin-luciferase systems are being explored for their potential in bioimaging, allowing scientists to visualize cellular processes in real time.

The remarkable versatility of luciferin-luciferase reactions extends beyond their use as analytical tools. Scientists are now engineering novel luciferin-luciferase systems with tailored properties. By modifying the structure of luciferin, researchers can create bioluminescent probes that emit light at specific wavelengths, enabling the simultaneous visualization of multiple cellular targets. Similarly, engineering luciferase enzymes can improve the brightness and stability of bioluminescent signals, opening up new possibilities for in vivo imaging and biosensing. The ongoing research into luciferin-luciferase reactions continues to illuminate our understanding of bioluminescence and its potential applications.

The exploration of luciferin-luciferase systems provides a foundational understanding of the diverse mechanisms employed by organisms to produce light. From the flashing signals of fireflies to the shimmering glow of marine organisms, these reactions underlie a remarkable array of bioluminescent phenomena. By delving into the molecular intricacies of these systems, we gain valuable insights into the evolutionary adaptations that have given rise to living light. Moreover, the study of luciferin-luciferase reactions has unlocked a wealth of applications in biotechnology, paving the way for innovative tools in research, diagnostics, and beyond. As we continue to unravel the secrets of these fascinating systems, we can anticipate even more exciting discoveries in the years to come.

3.2 Photoprotein Mechanisms

Photoproteins offer a captivating glimpse into the intricate world of bioluminescence, distinct from the more commonly known luciferin-luciferase systems. These remarkable proteins possess an inherent ability to emit light, a characteristic stemming from their unique structure and the tightly bound chromophore within. This chromophore, a molecule responsible for absorbing and emitting light, is intrinsically linked to the protein itself, creating a stable complex poised for luminescence. Unlike luciferase-based systems that require an external luciferin substrate, photoproteins hold the key to light production internally, ready to be unlocked under specific conditions. This pre-charged state distinguishes them and allows for a rapid, flash-like emission of light when triggered.

The best-studied example of a photoprotein is aequorin, isolated from the jellyfish Aequorea victoria. Aequorin's structure is a fascinating interplay of protein folds that encapsulate a coelenterazine molecule, the light-emitting chromophore. This coelenterazine, in its oxidized form, is already bound to the protein, along with oxygen, forming a stable peroxy-coelenterazine intermediate. This complex sits patiently, awaiting the arrival of a calcium ion. Upon binding calcium, a conformational change sweeps through the protein, catalyzing the oxidation of coelenterazine. This oxidation event releases energy in the form of a photon of blue light, a brilliant flash that illuminates the jellyfish's ethereal form. Other photoproteins have been discovered in various marine organisms, each exhibiting subtle variations in their structure and the color of light they emit. Obelin, found in the hydroid Obelia longicaudata, and thalassicolin, isolated from the jellyfish Thalassocalyce inconstans, are prime examples. These proteins, while sharing the fundamental mechanism of calcium-triggered luminescence, showcase the adaptability of photoprotein systems. Variations in the protein environment surrounding the coelenterazine influence the energy levels of the excited state, resulting in a spectrum of emitted light, from blue to green. Studying these variations provides valuable insights into the evolutionary tuning of bioluminescence.

The intricate relationship between the photoprotein structure and the emitted light color offers a rich area of exploration. Specific amino acid residues within the protein's active site play a crucial role in fine-tuning the emitted wavelength. By altering these residues through mutagenesis, researchers can manipulate the light color, effectively creating a palette of bioluminescent hues. This ability to engineer photoproteins has far-reaching implications for various fields, from bioimaging to biosensors.

The sensitivity of photoproteins to calcium ions makes them invaluable tools in cellular biology research. Aequorin, in particular, has been widely used as a calcium indicator in living cells. By introducing aequorin into cells, scientists can monitor intracellular calcium concentrations in real-time, gaining insights into fundamental cellular processes such as muscle contraction, nerve impulse transmission, and even programmed cell death. The rapid and highly specific response of aequorin to calcium fluctuations allows for precise measurements, even within the dynamic environment of a living cell.

Beyond cellular studies, photoproteins hold promise in broader biotechnological applications. Their inherent light-emitting capability opens doors to developing novel biosensors for detecting various analytes. By coupling photoproteins with specific binding domains that recognize target molecules, scientists can engineer biosensors that emit light upon binding, providing a readily detectable signal. These biosensors could be used for environmental monitoring, detecting pollutants or toxins, or in medical diagnostics, for example, detecting specific biomarkers of disease.

Further understanding of the mechanisms governing photoprotein function is crucial for unlocking their full potential. While significant progress has been made in elucidating the structure and bioluminescence mechanism of aequorin, other photoproteins remain less well-characterized. Investigating the intricacies of these systems will not only deepen our understanding of bioluminescence but also pave the way for developing innovative applications based on their remarkable properties. The ability to harness the light-emitting power of photoproteins offers a bright future for bioluminescence-based technologies.

The unique characteristic of photoproteins, their self-contained luminescent capacity, sets them apart within the broader landscape of bioluminescence. Their sensitivity to specific triggers like calcium ions and the ability to fine-tune their emitted light color through protein engineering makes them powerful tools for scientific exploration and technological innovation. As we delve deeper into the intricacies of photoprotein mechanisms, we unlock a world of possibilities for understanding and harnessing the magic of bioluminescence. This exploration promises to illuminate not only the fundamental processes of life but also to inspire new approaches to sensing, imaging, and manipulating biological systems.

3.3 Bacterial Bioluminescence

Bacterial bioluminescence offers a fascinating glimpse into the intricate world of light production in living organisms. Unlike the flash of a firefly or the shimmering glow of some fungi, bacterial light is often continuous, a steady beacon in the dark depths of the ocean or the moist recesses of a decaying log. This constant luminescence stems from the unique metabolic processes within these tiny organisms, processes intricately linked to their survival and interactions with their environment.

The key to bacterial bioluminescence lies within a complex chemical reaction involving two primary components: luciferase, the enzyme that catalyzes the reaction, and luciferin, the substrate that is oxidized to produce light. In bacteria, the luciferin is a reduced flavin mononucleotide, a molecule also involved in cellular respiration, highlighting the intertwined nature of light production and energy metabolism in these organisms. The bacterial luciferase enzyme, a heterodimer composed of two subunits designated and , facilitates the oxidation of the reduced flavin mononucleotide and a long-chain aldehyde, ultimately resulting in the emission of photons and the characteristic blue-green glow. Oxygen plays a critical role in this process, serving as the final electron acceptor and driving the reaction towards light production. This dependence on oxygen explains why many bioluminescent bacteria are found in aerobic environments or associated with hosts that provide access to oxygen-rich areas.

Diving deeper into the molecular mechanisms, we find that bacterial luciferase is encoded by the lux operon, a cluster of genes that work together to regulate and execute the bioluminescence process. The lux operon typically includes genes for both the and subunits of luciferase (luxA and luxB, respectively), as well as genes involved in the synthesis and recycling of the aldehyde substrate (luxC, luxD, and luxE). This organized genetic structure ensures the efficient production and control of all the necessary components for bioluminescence. The regulation of the lux operon itself is often influenced by population density, a phenomenon known as quorum sensing. Bacteria use small signaling molecules, autoinducers, to communicate with each other. When the bacterial population reaches a certain threshold, the concentration of autoinducers triggers the activation of the lux operon, leading to a synchronized burst of bioluminescence across the entire population.

The spectral properties of bacterial bioluminescence are also noteworthy. The emitted light typically falls within the blue-green spectrum, with a peak wavelength around 490 nanometers. This wavelength is particularly well-suited for transmission through seawater, making it an effective communication signal in marine environments. The specific spectral characteristics can vary slightly depending on the bacterial species and environmental factors such as temperature and pH. These variations can provide subtle nuances in the light signal, potentially conveying different information within bacterial communities.

Examining the ecological roles of bacterial bioluminescence reveals a variety of fascinating adaptations. In symbiotic relationships with marine animals, bacteria residing in specialized light organs provide a source of illumination for their hosts. This light can serve various purposes, from attracting prey to camouflaging against predators or even facilitating communication. For free-living bacteria, bioluminescence may play a role in attracting or deterring other organisms. In some cases, the light may act as a lure, attracting small crustaceans that inadvertently ingest the bacteria. In other instances, the light may serve as a warning signal, deterring potential predators.

Understanding the intricate workings of bacterial bioluminescence has broader implications beyond the realm of basic biology. The lux operon and its associated proteins have become valuable tools in biotechnology and bioengineering. Researchers have harnessed the power of bacterial bioluminescence to develop biosensors for detecting pollutants, monitoring gene expression, and even tracking the spread of infectious diseases. The sensitivity and specificity of the lux system make it an ideal platform for developing novel diagnostic and imaging techniques. Furthermore, the unique spectral characteristics of bacterial bioluminescence have inspired the development of new light-emitting materials and devices with potential applications in diverse fields, including optoelectronics and nanotechnology.

The continuous luminescence of bacteria, driven by their unique metabolic processes and regulated by sophisticated genetic mechanisms, exemplifies the remarkable diversity and adaptability of life. By delving into the intricacies of bacterial bioluminescence, we gain not only a deeper understanding of the natural world but also valuable insights that can be translated into practical applications across a range of scientific disciplines. The ongoing research into bacterial bioluminescence promises to unveil further secrets of this fascinating phenomenon and unlock even more innovative applications in the years to come. The exploration of these light-emitting bacteria sheds light, quite literally, on the complex interplay between biology, chemistry, and the environment.

3.4 Fungal Bioluminescence

The ethereal glow emanating from certain fungi has captivated humans for centuries. From the eerie foxfire illuminating forest floors to the ghostly shimmer of decaying wood, fungal bioluminescence represents a unique and fascinating adaptation. Unlike the flash of fireflies or the constant glow of some deep-sea creatures, fungal bioluminescence is often continuous, creating a persistent, otherworldly light. This chapter delves into the intricacies of this phenomenon, exploring the biochemical mechanisms, ecological roles, and evolutionary origins of light production in fungi.

Currently, around 100 species of bioluminescent fungi are known, primarily found in tropical and subtropical regions. These luminous fungi belong to the order Agaricales, the same order that includes common mushrooms. The majority of these species are from the genera Mycena, Omphalotus, Panellus, and Pleurotus. Interestingly, bioluminescent fungi often grow on decaying wood, playing a crucial role in nutrient cycling within forest ecosystems. Their light, however, is not merely a byproduct of decomposition.

The biochemical process behind fungal bioluminescence shares striking similarities with that of fireflies and other bioluminescent organisms. It involves a light-emitting molecule called luciferin, an enzyme called luciferase, and the presence of oxygen. In fungi, the specific luciferin molecule responsible for light emission was long a mystery. Recent research, however, has identified hispidin, a fungal metabolite, as a key compound in the light production pathway. Hispidin undergoes an enzymatic oxidation catalyzed by luciferase, resulting in the emission of light, primarily in the green-yellow spectrum, with wavelengths typically peaking around 520-530 nanometers. This creates the characteristic ghostly green glow often observed in these fungi.

Why do fungi glow? Several hypotheses have been proposed to explain the ecological significance of fungal bioluminescence. One prominent theory suggests that the light attracts insects and other arthropods. These attracted invertebrates may play a role in dispersing fungal spores, similar to how pollinators spread pollen in flowering plants. Imagine a tiny fungus, its delicate fruiting body glowing softly in the darkness, drawing the attention of a foraging beetle. As the beetle crawls over the fungus, it inadvertently picks up spores, carrying them to new locations and aiding in the fungus's propagation.

Another hypothesis posits that bioluminescence serves as a warning signal, deterring animals from consuming the fungus. This "warning glow" could indicate the presence of toxins or other unpalatable compounds within the fungal tissue. The constant light emission, unlike the intermittent flashes of fireflies, might reinforce this warning, ensuring that potential predators learn to associate the glow with an unpleasant experience.

Furthermore, some researchers suggest that bioluminescence might have a physiological role, acting as an antioxidant mechanism to protect the fungus from oxidative stress during the decomposition process. The enzymatic reaction that produces light could potentially neutralize harmful reactive oxygen species generated during metabolism. While this hypothesis requires further investigation, it adds another layer of complexity to the understanding of fungal bioluminescence.

The evolutionary origins of fungal bioluminescence remain an open question. While the specific evolutionary pressures that led to the development of light production in fungi are not fully understood, the widespread occurrence of this phenomenon across different fungal lineages suggests that it has evolved independently multiple times. This convergent evolution highlights the adaptive significance of bioluminescence in various ecological contexts. Further research into the phylogenetic relationships of bioluminescent fungi could shed light on the evolutionary history of this captivating trait.

The study of fungal bioluminescence offers a unique window into the complex interplay between organisms and their environment. By deciphering the biochemical mechanisms, ecological roles, and evolutionary history of this fascinating phenomenon, researchers can gain a deeper understanding of the diverse adaptations that have shaped life on Earth. From the dimly lit forest floor to the laboratory bench, the ghostly glow of bioluminescent fungi continues to illuminate the path towards new discoveries. Ongoing research into the genetic basis of fungal bioluminescence could even lead to biotechnological applications, such as the development of bioluminescent markers for various scientific purposes. The possibilities are as vast and intriguing as the mysterious glow itself.

3.5 Marine Bioluminescence

The ocean's depths, a realm of perpetual twilight and impenetrable darkness, harbor a breathtaking spectacle: bioluminescence. This living light, produced by a myriad of marine organisms, transforms the underwater world into a mesmerizing display of flashing, shimmering, and glowing displays. From the microscopic to the massive, bioluminescence serves a variety of functions, making it a crucial aspect of marine ecosystems. Understanding its diverse roles is key to appreciating the complex interplay of life in the ocean.

One of the primary functions of bioluminescence in the marine environment is defense. Many organisms utilize light flashes as a startling deterrent against potential predators. Imagine a small crustacean, suddenly illuminated by its own bioluminescent burst, momentarily blinding and confusing an approaching fish. This brief respite can be the difference between life and death, allowing the crustacean to escape. Some species even employ a strategy called "counterillumination," using ventrally located light organs to match the faint downwelling sunlight, effectively camouflaging their silhouettes against the brighter surface waters and making them invisible to predators lurking below.

Communication represents another significant role of bioluminescence. Certain species of fish, for example, use intricate light patterns to attract mates in the darkness of the deep sea. These flashing signals, unique to each species, ensure reproductive success by helping individuals identify compatible partners. Similarly, some deep-sea organisms utilize bioluminescent signals for schooling behaviors, enabling them to maintain cohesive groups in the absence of ambient light. This coordinated movement can enhance foraging efficiency and provide collective defense against predators.

Luring prey is a particularly cunning application of bioluminescence. The anglerfish, a deep-sea predator, dangles a bioluminescent lure in front of its mouth, attracting smaller fish and invertebrates who mistake the glowing appendage for a meal. When the unsuspecting prey comes close enough, the anglerfish swiftly engulfs it. This ingenious strategy highlights the evolutionary advantage conferred by bioluminescence in the challenging environment of the deep sea.

Beyond these primary functions, bioluminescence plays a surprising role in symbiotic relationships. Certain species of squid, for instance, harbor bioluminescent bacteria within specialized light organs. These bacteria provide a continuous glow that helps the squid blend in with the ambient light filtering down from the surface, offering camouflage against predators. In return, the bacteria receive a safe haven and a steady supply of nutrients from the squid. This mutually beneficial partnership demonstrates the intricate ecological connections facilitated by bioluminescence.

The diversity of bioluminescent organisms in the marine environment is staggering. From single-celled dinoflagellates that create sparkling waves along the shoreline to the deep-sea jellyfish with pulsating bells of light, the ocean teems with organisms capable of producing their own illumination. Each species has evolved unique adaptations, tailoring its bioluminescent displays to its specific ecological niche. This remarkable diversity underscores the evolutionary success of bioluminescence as a survival strategy in the marine world.

Exploring the mechanisms behind marine bioluminescence reveals a fascinating interplay of chemistry and biology. The light production process typically involves a light-emitting molecule called luciferin and an enzyme called luciferase. When luciferin reacts with oxygen in the presence of luciferase, it releases energy in the form of light. Different organisms utilize different types of luciferins and luciferases, leading to a wide range of colors and intensities of bioluminescence. Understanding these chemical pathways is crucial for unraveling the complexities of this remarkable phenomenon.

The study of marine bioluminescence continues to yield new insights into the ecology and evolution of life in the ocean. Researchers are developing innovative technologies to explore the deep sea, capturing stunning images and videos of bioluminescent displays in their natural habitat. These explorations are revealing the vast, unexplored biodiversity of the deep ocean and highlighting the importance of conserving these unique ecosystems. As we delve deeper into the mysteries of the deep, marine bioluminescence stands as a testament to the ingenuity and adaptability of life on Earth. Further research into this fascinating phenomenon promises to unlock even more secrets of the underwater world, illuminating the complex interactions between organisms and their environment. By understanding the intricacies of marine bioluminescence, we can gain a deeper appreciation for the interconnectedness of life on our planet and the remarkable adaptations that have allowed it to thrive in even the most extreme environments. The ongoing study of bioluminescence continues to inspire awe and wonder, reminding us of the vast, unexplored frontiers that still exist within our oceans.

4 Spectral Analysis

Understanding the spectral properties of light emitted by bioluminescent organisms is crucial for establishing and exploring the potential synchronicity between them and artificial organisms. This chapter will delve into the practical techniques used to measure and analyze these spectral properties. The primary focus will be on acquiring emission spectra, which are graphical representations of light intensity as a function of wavelength. These spectra provide a unique fingerprint of the emitted light, allowing for detailed comparisons between different organisms and artificial constructs.

Spectral measurement typically begins with the careful collection of emitted light. For bioluminescent organisms, this can involve specialized equipment designed for in situ measurements in various environments, from the depths of the ocean to terrestrial ecosystems. These tools minimize disturbance to the organisms and their natural light emission patterns. For artificial systems, the process is often simpler, involving controlled laboratory setups where the emitted light can be directly channeled to the measuring instrument. This allows for precise control over the experimental parameters. Regardless of the source, the collected light is then directed to a spectrometer.

A spectrometer is the heart of spectral analysis. This instrument separates the incoming light into its constituent wavelengths using either a prism or a diffraction grating. Prisms rely on the phenomenon of refraction, where different wavelengths of light bend at different angles as they pass through the prism material. Diffraction gratings, on the other hand, utilize the principle of diffraction, where light waves spread out as they pass through a series of closely spaced slits. Both methods effectively disperse the light, allowing for the measurement of its intensity at each wavelength. Modern spectrometers often employ sophisticated detectors, such as photomultiplier tubes or charge-coupled devices (CCDs), to accurately quantify the intensity of light at each separated wavelength.

The resulting data is then processed to generate an emission spectrum. This spectrum provides a wealth of information about the emitted light. The peak wavelength, corresponding to the color perceived by the human eye, is a key characteristic. The width of the spectrum, also known as the bandwidth, indicates the range of wavelengths present in the emitted light. The overall shape of the spectrum can reveal intricate details about the underlying light-emitting mechanisms. For instance, a narrow spectrum suggests a highly specific emission process, while a broader spectrum indicates a more complex process involving multiple energy transitions.

Analyzing emission spectra involves comparing and contrasting the spectral characteristics of different light sources. By comparing the emission spectra of a bioluminescent organism with that of an artificial light source, researchers can assess the degree of spectral overlap. This overlap is a key factor in determining the potential for synchronicity between the two systems. A significant overlap suggests that the two light sources share similar spectral properties, increasing the likelihood of interaction and potential synchronization. Conversely, a lack of overlap suggests that the two light sources are spectrally distinct, reducing the probability of interaction.

Beyond simply comparing peak wavelengths, researchers also examine the entire spectral profile. Subtle differences in the shape and width of the spectra can provide insights into the underlying bioluminescent and artificial light-producing mechanisms. For instance, slight shifts in peak wavelength or changes in the spectral bandwidth can indicate variations in the chemical composition of the light-emitting molecules or differences in the physical environment surrounding the light source. These nuances can be crucial in understanding the subtleties of spectral synchronicity. Moreover, time-resolved spectral measurements, which capture the spectrum at different points in time, allow for the study of dynamic changes in light emission. This is particularly important for investigating phenomena like flashing or pulsating bioluminescence, where the spectral characteristics may vary over time.

Furthermore, spectral analysis extends to characterizing the absorption spectra of photoreceptors in organisms. Photoreceptors are specialized proteins that absorb light, triggering a cascade of biological processes. By understanding the absorption spectrum of a photoreceptor, researchers can determine which wavelengths of light are most effective in eliciting a response. This knowledge is essential for studying how organisms perceive and react to light signals, both from their own species and from artificial sources. Comparing the emission spectrum of a light source with the absorption spectrum of a photoreceptor provides insights into the potential for communication and interaction between the two.

The application of spectral analysis is paramount in understanding the potential for synchronicity between artificial and biological systems. By precisely measuring and comparing the emission spectra of artificial organisms and bioluminescent organisms, we can quantify the degree of spectral overlap and determine the potential for interaction. This detailed spectral information provides a foundation for investigating the intricate dynamics of synchronized light patterns, paving the way for exciting new discoveries in the field of bioluminescence and artificial life. This analysis contributes significantly to the growing body of knowledge regarding the interaction of light between natural and engineered systems. The nuances uncovered by spectral analysis, from subtle shifts in peak wavelength to dynamic changes in emission profiles, can unlock a deeper understanding of the intricate relationship between light and life.

4.1 Spectral Measurement Techniques

Accurate spectral measurements are fundamental to understanding the interplay between artificial and biological light emission. This chapter will delve into the specific techniques used to capture and analyze these emissions, providing the foundation for comparing and correlating spectral signatures. We'll explore the instruments, methodologies, and critical considerations necessary for achieving precise and meaningful results in spectral analysis. This information will enable researchers to effectively characterize the light produced by both artificial organisms and bioluminescent systems, paving the way for deeper insights into their interactions.

Spectrometers form the core of spectral measurements. These instruments disperse light into its constituent wavelengths, allowing for the quantification of intensity at each wavelength. Different types of spectrometers exist, each with its own advantages and limitations. Grating spectrometers, for instance, utilize diffraction gratings to separate light based on wavelength, offering high resolution and broad spectral coverage. Prism spectrometers, on the other hand, employ prisms for dispersion, providing high throughput but potentially lower resolution. The choice of spectrometer depends on the specific application and the spectral characteristics of the light source being investigated. Beyond the basic instrument, considerations like detector sensitivity, spectral range, and resolution must be carefully evaluated.

Beyond the hardware, understanding the practical aspects of spectral measurement is crucial. Sample preparation is a key factor, especially when dealing with biological samples. Extraction methods, purification protocols, and sample handling techniques can significantly influence the accuracy of spectral readings. Ensuring sample stability and minimizing contamination are essential for reliable results. For artificial organisms, considerations include the materials used for light emission, the power source, and the control mechanisms for light production. Standardizing these parameters is critical for comparability across experiments.

Calibration is another vital aspect of accurate spectral measurement. Spectrometers need to be regularly calibrated using known light sources to ensure accurate wavelength and intensity readings. This involves using calibrated lamps or other standardized light emitters to establish a reference point for measurements. Regular calibration helps to correct for any instrument drift or variations over time. Furthermore, understanding the potential sources of error in spectral measurements is essential for minimizing their impact. These errors can arise from instrument noise, stray light, or variations in sample properties. Employing appropriate controls and error correction methods can improve the reliability and accuracy of the data.

Specific measurement protocols need to be tailored to the nature of the light source. For continuous light emissions, steady-state measurements can be employed, capturing the spectral distribution over a specific time interval. However, for transient or pulsed bioluminescence, time-resolved spectroscopy is necessary. This technique allows for the capture of spectral changes over very short time scales, revealing the dynamics of light emission processes. These measurements are particularly valuable for studying the kinetics of bioluminescent reactions and the temporal characteristics of artificial light sources.

Data analysis and interpretation are integral parts of spectral analysis. Raw spectral data often require processing to remove background noise, correct for instrument artifacts, and extract meaningful information. Various software tools and algorithms are available for spectral analysis, enabling researchers to perform tasks such as peak identification, area integration, and spectral fitting. These tools facilitate the quantitative analysis of spectral data, allowing for the comparison of different light sources and the identification of spectral shifts or changes in intensity. Furthermore, statistical analysis can be used to determine the significance of observed differences and to draw conclusions about the relationship between spectral properties and other experimental variables.

Comparing the spectral output of artificial organisms to that of natural bioluminescent systems requires careful consideration of several factors. Firstly, ensuring that measurements are conducted under comparable conditions is crucial. Factors like temperature, pH, and the presence of cofactors can influence bioluminescence and should be carefully controlled. Secondly, normalizing spectral data is essential for meaningful comparisons. This involves expressing the intensity at each wavelength as a fraction of the total emitted intensity, allowing for direct comparison of spectral shapes regardless of absolute intensity differences.

Finally, understanding the limitations of spectral measurements is important. While spectral analysis provides valuable information about the color and intensity of light, it does not reveal the underlying molecular mechanisms of light production. Complementary techniques, such as fluorescence microscopy and protein analysis, can provide further insights into the biological processes involved in bioluminescence. For artificial organisms, characterizing the materials and electronic components responsible for light emission is crucial for understanding the spectral properties observed. Combining spectral measurements with other analytical methods provides a more comprehensive understanding of both artificial and natural light emission, leading to a deeper appreciation of their interplay. This multifaceted approach will continue to drive the exploration of spectral synchronicity and its diverse applications. As technologies advance, so too will our ability to probe the intricate details of light emission and unlock the full potential of this fascinating phenomenon.

4.2 Emission Spectra of Organisms

Organisms, from fireflies to deep-sea creatures, have evolved bioluminescence for various purposes – communication, attracting mates, or luring prey. Understanding the spectral characteristics of these emissions is crucial for deciphering the complex interplay between light and life. The emission spectrum, essentially a fingerprint of the emitted light, reveals the specific wavelengths or colors an organism produces. This spectrum is shaped by the unique molecular structure of the luciferin, the light-emitting molecule, and the luciferase, the enzyme that catalyzes the reaction. The environment also plays a role, influencing the spectral output through factors like temperature, pressure, and the presence of other molecules that can absorb or scatter light. By analyzing these spectra, we gain insight into the evolutionary adaptations and ecological roles of bioluminescence.

Consider the firefly, a familiar symbol of bioluminescence. Its characteristic yellow-green flash is precisely tuned to attract mates within its specific habitat. The spectrum of this flash is remarkably narrow, concentrated around 560 nanometers, maximizing visibility against the backdrop of the night sky. This contrasts sharply with the broader emission spectra observed in some deep-sea organisms, which often produce blue or green light. These broader spectra might be advantageous in the deep ocean's dim, light-scattered environment, where reaching potential mates or prey requires a wider range of wavelengths. Analyzing these spectral differences helps us understand how bioluminescent systems have adapted to their respective ecological niches.

Moving beyond single organisms, consider the fascinating case of bioluminescent bacteria. Certain species of bacteria form symbiotic relationships with marine animals, residing in specialized light organs. These bacteria often emit a continuous glow, rather than the discrete flashes of fireflies. Their emission spectra can vary depending on the host species and the specific type of bacteria involved. For instance, the Hawaiian bobtail squid harbors Vibrio fischeri bacteria, which produce a blue-green light that helps the squid camouflage itself against the moonlit sky. Analyzing the spectral output of these symbiotic systems reveals the intricate co-evolutionary dynamics between host and symbiont.

Furthermore, the emission spectra of bioluminescent fungi offer another intriguing perspective. These fungi, found in decaying wood or leaf litter, emit a continuous greenish light. The exact function of fungal bioluminescence remains a topic of ongoing research, with hypotheses ranging from attracting insects to aid spore dispersal to deterring herbivores. Analyzing the spectral characteristics of these fungi and comparing them to other bioluminescent organisms can provide clues about the evolutionary origins and ecological significance of fungal light emission.

Now, consider the challenges of measuring and analyzing these diverse emission spectra. Specialized equipment is required to capture and quantify the faint light emitted by these organisms. Spectrometers, coupled with sensitive detectors, are used to precisely measure the intensity of light at different wavelengths. These instruments can be deployed in the field, allowing researchers to study bioluminescence in its natural context, or in laboratory settings, where environmental conditions can be carefully controlled. Data analysis techniques, such as spectral decomposition and curve fitting, are employed to extract meaningful information from the measured spectra.

In the field, researchers often use portable spectrometers to analyze the light emitted by organisms in their natural habitats. These devices allow for real-time measurements, capturing the dynamic changes in emission spectra that can occur in response to environmental stimuli or behavioral interactions. For instance, researchers have used portable spectrometers to study the flashing patterns of fireflies, revealing the intricate signaling mechanisms involved in mate attraction.

In laboratory settings, more sophisticated spectrometers and controlled environments allow for precise measurements and manipulation of variables. Researchers can isolate individual organisms or cultures to study their intrinsic spectral properties, eliminating the complexities of natural environments. This approach allows for detailed analysis of the underlying biochemical mechanisms of bioluminescence and the factors that influence spectral output.

Furthermore, advanced imaging techniques, such as hyperspectral imaging, provide spatially resolved spectral information. These techniques capture the emission spectra at multiple points across an image, revealing the spatial distribution of different bioluminescent components within an organism or a community. Such information is invaluable for understanding the complex interplay between bioluminescence and anatomical structures.

By combining these diverse approaches, researchers are gaining a deeper understanding of the fascinating world of bioluminescence. The emission spectra of organisms provide crucial insights into the evolutionary adaptations, ecological roles, and underlying biochemical mechanisms of this remarkable natural phenomenon. This knowledge not only expands our understanding of the natural world but also inspires the development of new technologies based on the principles of bioluminescence.

5 Synchronicity Phenomena

The convergence of engineered light and naturally occurring bioluminescence presents a fascinating arena for exploring synchronization. This phenomenon, where distinct entities exhibit coordinated behavior, manifests in various forms within the realm of light emission. Consider the fireflies of Southeast Asia, their individual flashes merging into a mesmerizing, collective pulse across the mangrove trees. This striking display of emergent synchronization, where independent oscillators lock into a shared rhythm, serves as a compelling natural model. We see similar patterns emerging in certain artificial systems designed to mimic bioluminescent properties.

Emergent synchronization relies on the subtle interplay of individual components. Imagine a network of artificial organisms, each programmed with a simple light-emitting algorithm. Initially, their flashes might appear random, lacking any discernible pattern. However, by introducing a mechanism for local interaction – perhaps by allowing each organism to respond to the light intensity of its neighbors – we can observe a remarkable transformation. As the system evolves, individual flashes begin to align, coalescing into a coordinated, rhythmic display. This transition from disorder to order is a hallmark of emergent synchronization. It highlights the power of local interactions to generate global patterns.

Coupled oscillators provide another framework for understanding synchronized behavior. Unlike emergent synchronization, which arises from indirect interactions, coupled oscillators involve direct physical or informational links between entities. Think of two pendulums mounted on a common base. Even if their initial swings are out of phase, the subtle vibrations transmitted through the base will eventually cause them to synchronize. In the context of bioluminescence, we can envision artificial organisms coupled through optical channels. The light emitted by one organism directly influences the emission of another, leading to synchronized oscillations in light intensity. The strength and nature of this coupling determine the resulting patterns, from simple in-phase oscillations to complex, multi-frequency rhythms.

Phase synchronization, a more subtle form of coordination, focuses on the temporal relationship between oscillating signals. Imagine two metronomes ticking at slightly different rates. While their frequencies may not perfectly align, their phases – the specific points within their respective cycles – can become synchronized. In the case of bioluminescent systems, phase synchronization can manifest as a consistent time lag between the light pulses of different organisms or between an artificial organism and a natural bioluminescent source. This type of synchronization may not be as visually striking as perfect frequency locking, but it nevertheless represents a significant form of coordination. It can reveal underlying connections and information flow between seemingly independent entities.

Spectral entrainment, a less explored but equally intriguing phenomenon, involves the adjustment of an oscillator's frequency to match an external signal. Imagine a radio receiver tuning into a specific broadcast frequency. The receiver's internal circuitry adjusts itself to resonate with the incoming signal, allowing it to be amplified and decoded. In the context of bioluminescence, spectral entrainment could occur when an artificial organism adjusts its emission spectrum to match that of a natural bioluminescent source. This dynamic interplay between emitted and received light has the potential to create intricate patterns of synchronized bioluminescence, potentially leading to novel forms of interspecies communication or environmental sensing. Exploring the mechanisms and implications of spectral entrainment holds great promise for future research.

The implications of these synchronicity phenomena extend beyond mere scientific curiosity. They pave the way for innovative applications in diverse fields. Consider the design of bio-inspired robots equipped with artificial bioluminescent systems. By leveraging the principles of synchronized light emission, these robots could communicate with each other underwater, coordinating their movements and actions in challenging environments. Imagine swarms of these robots exploring the deep ocean, their synchronized flashes illuminating the abyss and revealing its hidden secrets. Similarly, synchronized bioluminescence could revolutionize environmental monitoring. By engineering artificial organisms that respond to specific pollutants with changes in their light emission patterns, we can create sensitive and dynamic sensors for detecting environmental contamination. The synchronized flashes of these organisms could serve as an early warning system, alerting us to the presence of harmful substances.

The future of this field lies in continued exploration of the interplay between artificial and natural bioluminescence. By understanding the principles governing synchronicity phenomena, we can unlock new possibilities for bio-inspired technologies, environmental monitoring, and even interspecies communication. The subtle dance of light, whether generated by a firefly or an engineered device, holds immense potential for illuminating the mysteries of the natural world and inspiring groundbreaking innovations. The journey of discovery is just beginning, and the potential rewards are truly illuminating.

5.1 Emergent Synchronization

Synchronization, a ubiquitous phenomenon in nature, manifests spectacularly in the coordinated flashing of fireflies. This emergent behavior, where individual oscillators spontaneously align their rhythms, offers intriguing parallels to the potential interactions between artificial organisms and bioluminescent systems. Consider a swarm of bioluminescent artificial organisms engineered to emit light in similar spectral regions as their natural counterparts. Under the right conditions, these artificial organisms could synchronize their light emissions, creating a mesmerizing display akin to a firefly spectacle. The emergence of such synchronous behavior, however, isn't simply a matter of proximity. It's a complex interplay of individual oscillator properties and the coupling mechanisms that link them.

Imagine each artificial organism as a tiny oscillator, pulsing with light at a specific frequency. In isolation, these pulses might be random and uncoordinated. But introduce a coupling mechanism – perhaps a shared light field or a chemical signal – and the oscillators begin to influence each other. This interaction can lead to a gradual alignment of their phases, ultimately resulting in synchronous flashing. The strength of the coupling, the natural frequencies of the oscillators, and the presence of noise or external perturbations all play crucial roles in determining the final synchronized state. A weak coupling might only induce partial synchronization, where some oscillators align while others remain out of sync. Conversely, a strong coupling could lead to rapid and robust synchronization across the entire swarm.

Now, consider the implications of introducing these artificial organisms into a natural bioluminescent environment. Could they synchronize with naturally occurring organisms, creating a hybrid light show? The answer hinges on the spectral overlap between the artificial and natural light emissions. If the spectral peaks align closely, the natural bioluminescent organisms might perceive the artificial light as a signal from their own species. This could potentially entrain their flashing rhythms, drawing them into a synchronized dance with the artificial organisms. However, if the spectral mismatch is significant, the natural organisms might ignore the artificial light altogether, precluding any form of interaction.

The specific mechanisms driving synchronization in these hybrid systems can be diverse. In some cases, the coupling might be purely photonic, mediated by the exchange of photons between the artificial and natural light sources. The intensity and spectral composition of the artificial light could directly influence the firing rate of the natural bioluminescent organisms, driving them towards synchronization. In other scenarios, the coupling could be chemical, involving the release of signaling molecules that affect both the artificial and natural organisms. For example, some marine organisms use chemical signals to trigger bioluminescence, and similar mechanisms could be engineered into artificial organisms to facilitate synchronization.

Achieving robust synchronization requires careful tuning of several parameters. The intensity and pulse frequency of the artificial light emissions need to be within a certain range to effectively influence the natural organisms. The spatial distribution of the artificial organisms also plays a crucial role. A dense cluster of artificial organisms might exert a stronger influence than a dispersed group. Furthermore, environmental factors such as temperature, water currents, and background light levels can affect the bioluminescent activity of natural organisms, impacting the likelihood of synchronization.

Investigating the dynamics of these hybrid systems presents exciting research opportunities. By manipulating the properties of the artificial organisms and observing their interactions with natural bioluminescent communities, we can gain deeper insights into the fundamental principles governing synchronization. Mathematical models can help us predict the conditions under which synchronization is most likely to occur and explore the different patterns that can emerge. Experimental studies, both in controlled laboratory settings and in the field, can provide empirical validation of these models and reveal the complexities of real-world interactions. Ultimately, understanding the interplay between artificial and natural bioluminescence could pave the way for novel applications in areas such as underwater communication, environmental monitoring, and bio-inspired robotics. Imagine swarms of synchronized artificial organisms illuminating the ocean depths, tracking pollution plumes, or even communicating with marine life through coordinated light signals. The potential is vast, and the journey of exploration has just begun.

5.2 Coupled Oscillations

The intricate dance of fireflies on a summer night, the rhythmic pulsing of jellyfish in the deep ocean—these captivating displays of synchronized bioluminescence hint at underlying principles governing coordinated behavior. This chapter delves into the concept of coupled oscillations, a fundamental mechanism behind many synchronization phenomena observed in both natural and artificial systems. We will explore how this concept applies to the interaction between artificial organisms and bioluminescent life, offering insights into the complex dynamics of light-based communication and coordination.

Consider two pendulums hanging from a common support, connected by a spring. When one pendulum is set in motion, its oscillations gradually transfer energy to the other through the connecting spring. Eventually, both pendulums swing in unison, demonstrating a simple form of coupled oscillations. This principle extends beyond mechanical systems to encompass a wide range of phenomena, including the synchronous flashing of fireflies and the coordinated beating of heart cells. In our context, coupled oscillations can explain how artificial organisms, engineered to emit light, can synchronize their emissions with naturally bioluminescent organisms.

The mathematical framework for describing coupled oscillations often involves systems of differential equations that capture the interplay between individual oscillators and their coupling. These equations consider factors such as the natural frequency of each oscillator, the strength of the coupling, and any damping or external forces present. By analyzing these equations, we can predict the conditions under which synchronization is likely to occur and the characteristics of the resulting synchronized state. For instance, the Kuramoto model, a cornerstone of synchronization studies, describes the behavior of a large population of coupled oscillators, revealing how collective synchronization emerges from individual interactions. Adapting such models to the specific characteristics of artificial and biological light emitters provides a powerful tool for understanding and predicting their synchronized behavior.

The coupling between artificial and biological oscillators can take various forms. Direct physical coupling, akin to the spring connecting the pendulums, might involve a shared medium that transmits light energy between the organisms. For example, imagine an artificial organism embedded in a colony of bioluminescent bacteria. The emitted light from the bacteria could influence the light production of the artificial organism, leading to synchronized oscillations. Alternatively, indirect coupling could occur through environmental cues, such as changes in temperature or chemical concentrations, that affect both artificial and biological oscillators. This could be analogous to two metronomes placed on a common surface, synchronizing their ticking through vibrations transmitted through the surface.

Achieving robust synchronization between artificial and biological systems requires careful consideration of several factors. The spectral overlap between the emitted light of the artificial organism and the sensitivity of the bioluminescent organism's photoreceptors plays a crucial role. If the artificial light falls outside the spectral range detectable by the biological organism, effective coupling is unlikely. Furthermore, the temporal characteristics of the light emissions, such as the frequency and duration of pulses, must be compatible to allow for synchronization. Imagine trying to synchronize a slow, rhythmic drumbeat with a rapid, staccato flute melody – the mismatch in tempos would prevent any coordinated rhythm from emerging.

The ability to control and manipulate coupled oscillations opens up exciting possibilities for various applications. By designing artificial organisms that can synchronize with naturally bioluminescent species, we can gain deeper insights into the communication and coordination strategies employed by these organisms. This could lead to new approaches for studying and interacting with complex biological systems. For example, imagine deploying a swarm of artificial organisms programmed to synchronize with the flashing of fireflies, allowing researchers to monitor the fireflies' behavior and potentially even influence their communication patterns.

Moreover, coupled oscillations could be exploited for developing novel bio-inspired technologies. Synchronized light emissions could be used for underwater communication, environmental monitoring, or even artistic expression. Imagine a network of underwater sensors, mimicking the synchronized flashing of deep-sea creatures, transmitting data through coordinated light pulses. Or envision an art installation where artificial organisms synchronize their light emissions with bioluminescent algae, creating a mesmerizing display of pulsating light and color.

Further research into coupled oscillations in the context of artificial organisms and bioluminescence promises to reveal deeper insights into the fundamental principles governing synchronization in nature. By exploring the interplay between these two domains, we can unlock new possibilities for understanding, interacting with, and even harnessing the power of biological light. This knowledge will pave the way for developing innovative technologies and addressing complex challenges across diverse fields, from robotics to environmental science to art.

5.3 Phase Synchronization

Phase synchronization represents a specific type of emergent behavior observed when two or more oscillating systems interact. It's not about the oscillators becoming identical in frequency; rather, it's about their phases locking into a consistent relationship. Imagine two fireflies flashing, not necessarily at the same rate, but with a predictable time delay between their flashes. This consistent phase relationship, regardless of the individual frequencies, is the hallmark of phase synchronization. This phenomenon plays a crucial role in understanding how complex systems, both biological and artificial, can achieve coordinated behavior.

The mathematical framework for understanding phase synchronization involves analyzing the phase difference between the oscillating systems. If this phase difference remains constant over time, the systems are considered phase-synchronized. This constancy can be represented mathematically using equations that describe the evolution of the phase difference. These equations often involve coupling terms that represent the interaction between the oscillators. The strength of this coupling influences the likelihood and stability of phase synchronization. Weak coupling might lead to intermittent synchronization, while stronger coupling can result in robust, persistent phase locking.

In the context of bioluminescent organisms, phase synchronization can manifest in striking visual displays. Consider schools of fish flashing in unison, creating mesmerizing underwater light shows. These coordinated displays are often attributed to phase synchronization among the individual fish's bioluminescent systems. Each fish acts as an oscillator, and their interactions, perhaps mediated by the surrounding water or through direct visual cues, lead to the emergence of synchronized flashing patterns. This synchronization can serve various biological functions, from confusing predators to attracting mates.

When considering artificial organisms designed to mimic bioluminescence, achieving phase synchronization introduces a new layer of complexity. We must not only replicate the light-emitting properties of biological systems but also engineer the appropriate interactions that allow for synchronization. This might involve designing communication pathways between artificial organisms, allowing them to exchange information about their phase and frequency. It could also entail creating environmental cues that influence the oscillatory behavior of the artificial organisms, mimicking the role of the environment in natural bioluminescent systems.

Achieving phase synchronization in artificial systems can open doors to novel applications. Imagine swarms of bioluminescent robots coordinating their light emission to illuminate underwater environments or to signal each other in complex search and rescue operations. Phase synchronization could also enable new forms of communication, allowing artificial organisms to transmit information through modulated light patterns, mimicking the complex signaling observed in some bioluminescent species. Moreover, studying phase synchronization in artificial systems can provide valuable insights into the underlying mechanisms driving synchronization in natural biological systems, furthering our understanding of collective behavior in the natural world.

The process of inducing phase synchronization in artificial organisms can involve different strategies depending on the specific application and the nature of the organisms themselves. One approach involves direct coupling, where the oscillators within the artificial organisms are physically or electronically linked. This direct connection ensures that information about phase and frequency is readily exchanged, facilitating the establishment of a stable phase relationship. Another strategy involves indirect coupling, where the organisms interact through a shared environment. Changes in the environment, such as light intensity or chemical concentrations, can influence the oscillatory behavior of the individual organisms, indirectly leading to synchronization.

A key challenge in achieving phase synchronization lies in overcoming noise and variability. In real-world systems, both natural and artificial, oscillators are subject to fluctuations and disturbances that can disrupt the delicate balance required for synchronization. Robust synchronization mechanisms must be able to withstand these perturbations and maintain a consistent phase relationship. This robustness can be achieved through various means, such as incorporating feedback loops that compensate for deviations from the desired phase or by designing oscillators with inherent stability against noise. The development of such robust synchronization mechanisms is crucial for realizing the full potential of bio-inspired technologies that rely on coordinated light emission.

Investigating the dynamics of phase synchronization in artificial organisms can also shed light on the evolutionary pressures that shaped bioluminescent communication in nature. By manipulating the parameters of the artificial systems, researchers can explore how different factors, such as communication range and environmental noise, influence the evolution of synchronization strategies. These studies can provide valuable insights into the selective advantages conferred by synchronized bioluminescence in various ecological contexts. Furthermore, understanding the principles of phase synchronization can inform the design of bio-inspired lighting systems, enabling the creation of dynamic and energy-efficient lighting solutions that mimic the coordinated displays observed in nature.

The study of phase synchronization in artificial organisms requires sophisticated tools and techniques. Precise control over the oscillatory behavior of the artificial organisms is essential, often involving advanced electronics and microfluidic devices. Analyzing the resulting light patterns requires sensitive photodetectors and specialized software for quantifying phase relationships and identifying synchronization patterns. Furthermore, computational modeling plays a crucial role in understanding the complex dynamics of interacting oscillators and in predicting the emergent behavior of synchronized systems. By combining experimental and computational approaches, researchers can gain a deeper understanding of the principles governing phase synchronization and unlock its potential for technological innovation.

5.4 Spectral Entrainment

Spectral entrainment, a fascinating phenomenon within the broader context of synchronization, describes the adjustment of one oscillator's rhythm to match the frequency of another, stronger oscillator. Imagine two pendulums on a slightly flexible beam; over time, their swings, even if initially different, will synchronize. This principle, observable in mechanical systems, extends to biological and artificial light-emitting systems, forming a core concept in our exploration of spectral synchronicity. In the realm of bioluminescence, fireflies flashing in unison demonstrate a compelling example of natural spectral entrainment, where individual signals coalesce into a mesmerizing collective display.

Consider the intricacies of this natural phenomenon. Fireflies, through precisely timed flashes, achieve synchronous communication, a testament to the power of entrainment in the natural world. Their internal biological clocks, influenced by the flashes of their neighbors, adjust to match the dominant rhythm, creating a breathtaking spectacle of coordinated light pulses. This natural orchestration, achieved through subtle variations in flash timing and intensity, highlights the complex interplay between individual oscillators and their environment.

Now, let's shift our focus to the artificial realm. Synthetic oscillators, designed to mimic the bioluminescent properties of organisms like fireflies, can also exhibit entrainment. Imagine a network of microscopic robots, each emitting light pulses at slightly different frequencies. By carefully manipulating the interaction parameters, such as the strength of coupling between oscillators and the sensitivity to external signals, we can induce spectral entrainment. This allows us to control the collective behavior of the artificial system, creating synchronized light patterns that mimic or even surpass the complexity of natural phenomena.

Achieving spectral entrainment in artificial systems requires a deep understanding of the underlying mechanisms. The coupling strength between oscillators plays a crucial role. Weak coupling may lead to slow or incomplete synchronization, while strong coupling can force rapid alignment but potentially suppress individual variations. The optimal coupling strength depends on the specific application and desired level of synchronization. For instance, in bio-inspired robotics, precise control over entrainment could enable coordinated movement and communication among a swarm of robots.

Furthermore, the spectral characteristics of the emitted light are critical. The overlap between the emission spectra of different oscillators influences the efficiency of entrainment. If the spectra are significantly different, the oscillators may be less susceptible to mutual influence, hindering synchronization. Therefore, careful selection of materials and design of light-emitting components is essential for achieving robust and predictable spectral entrainment.

Imagine a scenario where artificial organisms, embedded with precisely tuned light-emitting devices, are deployed in a marine environment. These devices, designed to mimic the spectral characteristics of bioluminescent marine organisms, could interact with and potentially influence the natural light patterns of the ecosystem. By carefully modulating the emission spectra and timing of the artificial organisms, we could explore the possibility of inducing spectral entrainment in the natural population, opening up new avenues for studying and interacting with marine ecosystems.

However, this potential also raises important ethical considerations. The ability to influence natural bioluminescent systems carries a responsibility to understand and mitigate any potential negative impacts. Careful experimentation and monitoring are crucial to ensure the responsible application of this technology. Furthermore, the development of clear ethical guidelines is necessary to navigate the complex interplay between artificial and natural systems.

The study of spectral entrainment provides valuable insights into the fundamental principles of synchronization and its role in both natural and artificial systems. By exploring the interplay between individual oscillators and their environment, we can gain a deeper understanding of how complex patterns emerge from simple interactions. This knowledge has far-reaching implications, from developing advanced bio-inspired technologies to unraveling the mysteries of bioluminescence in the natural world. Furthermore, the ability to manipulate and control spectral entrainment opens up exciting possibilities for creating novel light-based communication systems, designing sophisticated robotic swarms, and interacting with biological systems in unprecedented ways. As we delve deeper into the intricacies of spectral entrainment, we unveil a fascinating world of synchronized light, with the potential to transform our understanding of both the natural and artificial world. This is a journey of exploration, innovation, and responsible application, promising to illuminate the path towards a future where light plays an even more central role in our lives.

5.5 Applications of Synchronicity

The potential applications of synchronicity between artificial organisms and bioluminescent systems are vast and interdisciplinary, bridging the gap between biology, engineering, and computation. One promising area lies in bio-inspired robotics, where the principles of bioluminescent communication can be harnessed to create autonomous underwater vehicles capable of complex, coordinated behaviors. Imagine swarms of these robots, emitting synchronized flashes of light to communicate with each other, mapping the ocean floor, or tracking schools of fish with minimal disruption to the marine environment. This form of underwater communication, inspired by the firefly's mesmerizing light displays, could revolutionize oceanographic research and exploration, allowing for the collection of data in previously inaccessible environments. Furthermore, such bio-inspired robots could be deployed in search and rescue missions, using their synchronized bioluminescence to signal their location or illuminate disaster zones.

Beyond underwater exploration, synchronized bioluminescent artificial organisms offer exciting possibilities for environmental monitoring. These organisms, designed to react to specific environmental stimuli, could be deployed to detect pollutants in water bodies, monitor air quality, or even track the spread of invasive species. For example, an artificial organism could be engineered to emit a particular light signal when it encounters a specific toxin, providing real-time information about the presence and concentration of the pollutant. Such bio-sensors, distributed across a large area and synchronized to provide a comprehensive picture of environmental health, would be invaluable for conservation efforts and pollution control. This method allows for continuous, non-invasive monitoring, minimizing the need for manual sampling and laboratory analysis. The synchronized light output could also be used to visualize the spatial distribution of the pollutant, enabling rapid identification of contamination hotspots and targeted remediation efforts.

Another application of this technology lies in the development of novel display technologies. Imagine a screen composed of millions of microscopic artificial organisms, each capable of emitting light of varying intensities and colors. By controlling the synchronization of these organisms, incredibly detailed and dynamic displays could be created, surpassing the capabilities of current LCD and LED screens. Such bioluminescent displays would not only offer superior image quality but also consume significantly less energy, making them a more sustainable alternative to existing technologies. Further, the flexibility and biocompatibility of these organisms could open up exciting new avenues for wearable displays and implantable biomedical devices, potentially revolutionizing fields like personalized medicine and augmented reality.

In agriculture, synchronized bioluminescent artificial organisms could be employed to monitor plant health and optimize crop yields. These organisms, designed to interact with specific plant species, could detect early signs of disease or nutrient deficiency, allowing farmers to take preventative measures and minimize crop loss. The synchronized light emissions of these organisms could be used to create detailed maps of plant health across an entire field, providing valuable insights for precision agriculture techniques. This approach could lead to more efficient use of resources, reduced pesticide use, and ultimately, a more sustainable and productive agricultural system. The ability to monitor plant health at a microscopic level, coupled with the real-time feedback provided by synchronized bioluminescence, has the potential to transform farming practices.

Moreover, the principles of spectral synchronicity can be applied to develop new forms of optical communication. By modulating the light emissions of artificial organisms in a controlled manner, information can be encoded and transmitted using light signals. This approach could have applications in secure communication, underwater data transmission, and even interspecies communication. Imagine using light to communicate with genetically modified plants, receiving real-time feedback on their growth and health, or exchanging information with other organisms in a complex ecosystem. This opens up a whole new realm of possibilities for understanding and interacting with the natural world. The inherent security and low energy consumption of optical communication, combined with the potential for complex signaling patterns enabled by synchronized bioluminescence, offer significant advantages over traditional communication methods.

Finally, synchronized bioluminescence has the potential to revolutionize medical imaging and diagnostics. Artificial organisms emitting light at specific wavelengths could be used to target and illuminate specific tissues or organs within the body, providing real-time, non-invasive imaging capabilities. This approach could significantly improve the accuracy and speed of disease diagnosis, enabling earlier intervention and better treatment outcomes. Furthermore, synchronized bioluminescent organisms could be used to deliver drugs or therapies directly to diseased cells, minimizing side effects and maximizing treatment efficacy. This targeted approach to medicine, guided by the precise control of light emission and synchronization, holds immense promise for personalized healthcare and the development of new, more effective treatments. The biocompatibility and biodegradability of these organisms further enhance their potential for in vivo applications, offering a less invasive and more targeted approach to medical imaging and drug delivery.

6 Applications and Implications

The convergence of artificial organisms and bioluminescence offers unique opportunities in bio-inspired robotics. Imagine underwater robots, mimicking the camouflage and communication strategies of bioluminescent marine life. These robots, equipped with synthetic bioluminescent systems, could seamlessly blend into their environment, enabling unobtrusive observation of delicate ecosystems and elusive creatures. Further, by modulating their light emission, these robots could potentially communicate with marine organisms, opening new avenues for interspecies interaction and understanding. Beyond the ocean's depths, bio-inspired robots employing bioluminescence could navigate complex, light-deprived environments like caves or disaster zones, using their light for both navigation and communication. The development of soft, flexible materials that incorporate bioluminescent elements could lead to a new generation of robots capable of adapting to challenging terrains and performing delicate tasks. These robots could find applications in search and rescue missions, environmental monitoring, and even medical procedures. The potential of bio-inspired robotics based on bioluminescence is vast and remains largely unexplored, promising transformative advancements in various fields.

Beyond robotics, bioluminescent systems present powerful tools for environmental monitoring. Consider genetically engineered microorganisms that emit light in response to specific pollutants. Deployed in water bodies or soil, these microbial sensors could provide real-time, localized data on contamination levels, enabling rapid responses to environmental threats. Furthermore, the intensity and spectral characteristics of the emitted light could be tailored to indicate the type and concentration of the pollutant, creating a sophisticated and sensitive detection system. This approach offers a cost-effective and minimally invasive alternative to traditional monitoring methods, allowing for continuous, in situ assessment of environmental health. Moreover, bioluminescent sensors could be integrated into wearable devices or deployed in remote locations, providing crucial data for environmental protection and public health. By leveraging the sensitivity and specificity of bioluminescent reactions, we can develop advanced monitoring systems to safeguard our planet.

Bioluminescent organisms themselves can serve as valuable indicators of environmental health. Changes in their light emission patterns, intensity, or spectral composition can reflect alterations in water quality, temperature, or the presence of toxins. By monitoring these natural light signals, researchers can gain insights into the overall health of ecosystems and detect early warning signs of environmental stress. This approach, known as biomonitoring, offers a holistic perspective on ecosystem dynamics, complementing traditional methods that focus on individual parameters. For instance, the decline in bioluminescence intensity in a particular area could signal pollution or oxygen depletion, prompting timely interventions to mitigate the impact. Moreover, the spectral shifts in bioluminescence can reveal the presence of specific pollutants, providing valuable information for targeted remediation efforts. By harnessing the power of natural bioluminescence, we can gain a deeper understanding of complex environmental processes and protect vulnerable ecosystems.

The integration of artificial organisms with natural bioluminescent systems opens up exciting possibilities for studying and manipulating biological processes. Imagine artificial cells engineered to interact with bioluminescent bacteria, exchanging signals and influencing their light emission patterns. This could provide a unique platform for investigating interspecies communication and understanding the complex interplay between different organisms. Furthermore, by manipulating the spectral properties of artificial light sources, researchers could potentially control the behavior of bioluminescent organisms, opening up new avenues for bio-control and bio-remediation. For instance, artificial light could be used to attract or repel specific bioluminescent species, influencing their distribution and interactions within an ecosystem. This approach could have implications for controlling harmful algal blooms, managing invasive species, or even enhancing the bioluminescent properties of certain organisms for specific applications. The integration of artificial and natural bioluminescent systems offers a powerful toolkit for manipulating and understanding the living world.

The future of bioluminescence and its integration with artificial organisms is brimming with possibilities. Further research into the genetic mechanisms underlying bioluminescence could lead to the development of novel light-emitting materials and devices. These advances could revolutionize lighting technology, creating more energy-efficient and environmentally friendly light sources. Imagine cities illuminated by bioluminescent trees or buildings coated with bioluminescent paints, transforming urban landscapes and reducing our reliance on traditional lighting systems. Furthermore, the development of portable bioluminescent devices could provide safe and sustainable light sources for remote areas or disaster zones, empowering communities and enhancing resilience in the face of challenges. The exploration of bioluminescence and its integration with artificial systems holds immense potential for shaping a brighter and more sustainable future.

6.1 Bio-inspired Robotics

Nature has long served as a muse for engineers, and the ethereal glow of bioluminescent organisms offers a unique inspiration for robotics. This chapter explores how the principles of bioluminescence are being applied to create innovative robotic systems, focusing on the potential for underwater exploration, communication, and camouflage. We will delve into the challenges and advancements in mimicking bioluminescent processes and discuss the potential benefits of these bio-inspired designs.
Underwater environments present unique challenges for robotic exploration, including limited visibility and harsh conditions. Traditional lighting systems can disrupt the delicate balance of these ecosystems and are often ineffective due to scattering and absorption. Bioluminescent robotics offers a compelling alternative. By incorporating light-emitting components that mimic the subtle glow of marine organisms, robots can navigate the depths with minimal disruption, blending seamlessly with their surroundings. This approach allows for unobtrusive observation of marine life and the collection of valuable scientific data. Imagine a swarm of bioluminescent robots, pulsating with light, mapping the ocean floor or studying the behavior of deep-sea creatures without disturbing their natural habitat.
The communication capabilities of bioluminescent organisms also provide a valuable model for robotic design. Many marine species use light signals for mating, defense, and even hunting. Researchers are exploring how these natural communication strategies can be adapted to create underwater robotic networks. Bioluminescent robots could communicate with each other using coded light pulses, enabling coordinated movements and data sharing in environments where traditional communication methods are limited. This could revolutionize underwater exploration, allowing for complex tasks to be carried out by teams of robots working in concert. Picture a group of robots, each flashing a unique bioluminescent pattern, coordinating their search for a sunken

shipwreck or mapping the distribution of a specific species.
Camouflage is a crucial survival strategy for many bioluminescent organisms, allowing them to blend in with the ambient light of their environment. This ability to dynamically control light emission has inspired the development of adaptive camouflage systems for robots. By mimicking the bioluminescent patterns of their surroundings, robots can effectively disappear from view, potentially enabling covert operations or close-range observation of sensitive ecosystems. Consider a robot deployed in a coral reef, its surface dynamically adjusting its bioluminescence to match the shifting patterns of light and shadow, effectively becoming invisible to predators or prey. This opens exciting possibilities for studying elusive creatures and monitoring delicate environments without causing disturbance.
Replicating the complex chemical reactions that produce bioluminescence in living organisms is a significant challenge for robotics engineers. Current research focuses on developing synthetic materials that can mimic these processes, generating light with similar efficiency and spectral characteristics. Progress has been made in creating artificial luciferins and luciferases, the key components of many bioluminescent systems. These synthetic compounds can be incorporated into robotic platforms, providing a sustainable and controllable source of light. The development of flexible, biocompatible materials that can emit light is also a key area of research, paving the way for soft-bodied robots that can seamlessly integrate with their environment.
The development of bioluminescent robotics requires expertise from diverse fields, including biology, chemistry, materials science, and robotics engineering. Researchers are collaborating to overcome the challenges in mimicking natural bioluminescence and integrating these systems into functional robotic platforms. The potential benefits of this interdisciplinary approach are vast, ranging from improved underwater exploration and communication to novel applications in medicine, environmental monitoring, and even entertainment. Imagine tiny biolumi-

nescent robots navigating the human body, delivering drugs or performing microsurgery, or swarms of robots illuminating artistic installations with dynamic, living light displays.

While the field of bio-inspired robotics is still in its early stages, the progress made so far is promising. As our understanding of bioluminescence deepens and our ability to engineer bio-inspired materials improves, we can expect to see a growing range of applications for these innovative technologies. The future of bioluminescent robotics is bright, with the potential to illuminate new frontiers in exploration, communication, and our interaction with the natural world. The synergy between biology and robotics promises a future where technology seamlessly integrates with the living world, offering solutions that are both elegant and sustainable. These developments hold the key to unlocking new possibilities and understanding the intricate workings of nature's own light shows.

6.2 Environmental Monitoring

The increasing sophistication of artificial organisms capable of mimicking bioluminescence opens exciting possibilities for environmental monitoring. Imagine deploying swarms of these artificial organisms into delicate ecosystems – coral reefs, polluted waterways, or even agricultural fields. Their engineered bioluminescence, tuned to react to specific environmental parameters, could provide real-time data on water quality, pollutant levels, or plant health. This approach offers significant advantages over traditional methods, which often involve laborious sampling and laboratory analysis. Real-time data collection allows for immediate responses to environmental changes, potentially mitigating damage before it becomes irreversible.

Consider a scenario where artificial, bioluminescent organisms are released into a river system suspected of contamination. These organisms could be engineered to emit light at different wavelengths depending on the presence and concentration of specific pollutants. A shift in the overall light signature of the swarm, detectable by remote sensors, would immediately signal the presence and type of contamination. This rapid detection allows for immediate intervention, preventing further spread of pollutants and facilitating targeted cleanup efforts. The intensity of the emitted light could even be calibrated to correspond to the concentration of the pollutant, providing quantitative data for more precise assessments.

Another potential application lies in monitoring the health of coral reefs. Coral bleaching, a significant consequence of climate change, is often difficult to detect in its early stages. Artificial organisms, designed to integrate into the reef ecosystem and emit light in response to changes in water temperature, pH, or other relevant factors, could provide an early warning system for bleaching events. This early detection is crucial, as it allows for timely interventions like shading the reef or mitigating local stressors, potentially preventing widespread coral death. The data gathered by these artificial organisms could also contribute to a deeper understanding of the complex interplay of factors contributing to coral bleaching.

Moving beyond aquatic environments, these organisms could be utilized in terrestrial ecosystems as well. Imagine deploying bioluminescent artificial organisms in agricultural fields. These organisms could be engineered to react to soil conditions, emitting different light patterns in response to nutrient deficiencies, the presence of pests, or changes in water availability. This real-time monitoring would allow farmers to optimize irrigation, fertilization, and pest control strategies, leading to increased yields and reduced environmental impact through more precise resource management. The dynamic nature of this monitoring system provides constant feedback, allowing for adaptive adjustments to farming practices throughout the growing season.

The development of these environmental monitoring systems necessitates careful consideration of the organisms' design. The choice of the luciferin-luciferase system, the specific wavelengths of emitted light, and the organisms' physical characteristics must be tailored to the target environment and the parameters being monitored. For instance, in a murky river, longer wavelengths of light might be more effective for transmission and detection, whereas in a clear, shallow reef, shorter wavelengths might be preferable. Furthermore, the organisms' size, shape, and motility must be optimized for effective dispersal and integration into the target ecosystem.

Crucially, the bioluminescent response of these artificial organisms must be specific and sensitive to the target environmental parameter. False positives or insensitive responses could lead to inaccurate assessments and ineffective interventions. Careful calibration and rigorous testing in controlled environments are essential before deploying these organisms in real-world scenarios. The stability and longevity of the bioluminescent reaction also need to be considered, ensuring that the organisms can provide reliable data over extended periods. Finally, the potential ecological impact of introducing these artificial organisms must be thoroughly assessed and minimized to prevent unintended consequences. This includes careful selection of biocompatible materials and strategies for retrieval or degradation of the organisms at the end of their operational lifespan.

The potential of bioluminescent artificial organisms for environmental monitoring is vast. As the technology matures and our understanding of both bioluminescence and ecosystem dynamics deepens, these organisms will undoubtedly play an increasingly crucial role in protecting and preserving our planet. They offer a powerful new tool for understanding and responding to the complex challenges facing our environment, paving the way for a more sustainable future. This technology promises to transform how we monitor and manage our planet's precious resources.

By integrating principles of synthetic biology, materials science, and engineering, researchers can continue to refine and expand the capabilities of these artificial organisms, opening up new frontiers in environmental monitoring and remediation. The convergence of these disciplines is driving innovation in this field, leading to ever more sophisticated and effective tools for understanding and protecting our environment. The ongoing development of miniaturized sensors, advanced data processing algorithms, and robust bioluminescent systems will further enhance the power and versatility of this approach, allowing us to address increasingly complex environmental challenges.

7 Future Directions

The convergence of artificial organisms and bioluminescence presents a frontier ripe with possibilities. Consider the development of advanced artificial organisms capable of not just mimicking, but actively interacting with natural bioluminescent systems. Imagine artificial entities integrated into marine environments, communicating with bioluminescent organisms through precisely tuned light signals. This could revolutionize our understanding of complex marine ecosystems and potentially lead to new strategies for conservation and resource management. Such advancements would necessitate sophisticated sensory systems within the artificial organisms, allowing them to perceive and interpret the subtle nuances of bioluminescent communication in real-time. These artificial organisms could act as intermediaries, bridging the gap between human observation and the intricate language of light used by marine life.

Synthetic bioluminescence offers another exciting avenue. Moving beyond mimicking existing bioluminescent systems, we can envision designing entirely novel light-emitting mechanisms. Tailoring the spectral properties, intensity, and temporal patterns of artificial bioluminescence opens doors to applications beyond our current imagination. This could involve creating bioluminescent materials for medical imaging, developing self-powered lighting systems for sustainable urban environments, or even designing dynamic displays that interact with their surroundings. The ability to engineer bioluminescence at the molecular level would provide unprecedented control over light emission, offering a toolbox for innovation across diverse fields.

The prospect of interspecies communication through engineered bioluminescence is particularly compelling. By developing artificial organisms that can generate and interpret light signals specific to certain species, we could potentially establish two-way communication channels. This opens exciting possibilities for influencing animal behavior, studying complex ecological interactions, and even developing new forms of bio-inspired therapy. Imagine using artificial bioluminescence to guide migratory patterns, deter pests from crops, or even communicate with symbiotic bacteria within the human gut. The ethical implications of such interventions, however, must be carefully considered as we venture into this uncharted territory.

Developing artificial organisms capable of participating in the complex symphony of bioluminescent communication raises important ethical considerations. How do we ensure responsible use of this technology? What safeguards are needed to prevent unintended consequences for natural ecosystems? These questions require careful deliberation involving scientists, ethicists, and policymakers. Establishing clear guidelines for the development and deployment of artificial organisms with bioluminescent capabilities is crucial to mitigate potential risks and maximize the benefits for both humanity and the environment. We must consider the potential impact on existing ecological balances, the possibility of disrupting natural communication networks, and the long-term effects of introducing artificial entities into delicate ecosystems.

Numerous open questions remain to be explored. How can we improve the efficiency and longevity of synthetic bioluminescent systems? What are the long-term effects of introducing artificial organisms into natural ecosystems? Can we develop bio-inspired sensors that rival the sensitivity and specificity of biological light detection systems? These questions represent just a starting point, and as we delve deeper into the intersection of artificial organisms and bioluminescence, new challenges and opportunities will undoubtedly emerge. Further research is needed to understand the fundamental principles governing bioluminescent communication, to develop robust and adaptable artificial organisms, and to establish ethical frameworks that ensure responsible innovation in this rapidly evolving field. The potential rewards, however, are immense, promising transformative advances in our understanding of life, light, and the intricate connections between them. As we continue to push the boundaries of scientific inquiry, the interplay of artificial and natural light may illuminate pathways to a more sustainable and interconnected future.

7.1 Advanced Artificial Organisms

The convergence of advanced computing, materials science, and synthetic biology is paving the way for a new generation of artificial organisms. These are not merely simulations running on a computer, but physically embodied entities capable of exhibiting complex behaviors, including bioluminescence, in the real world. Imagine swarms of bioluminescent artificial organisms interacting with their environment, responding to stimuli, and even communicating with natural organisms through carefully orchestrated light signals. This isn't science fiction, but a tangible future rapidly approaching.

One crucial aspect of these advanced artificial organisms is their ability to adapt and learn. Early artificial life forms were programmed with fixed behaviors. However, future generations will incorporate learning algorithms, enabling them to modify their light emission patterns based on experience. This adaptability opens exciting possibilities for applications like environmental monitoring, where these organisms could adjust their bioluminescence to indicate the presence of pollutants or changes in environmental conditions. Consider a network of these organisms deployed in a polluted waterway, their light pulsing at different frequencies to reflect varying levels of contamination, providing real-time data for remediation efforts.

The physical construction of these advanced artificial organisms presents a unique set of challenges. Researchers are exploring innovative materials that mimic the flexibility and responsiveness of biological tissues. These materials could be combined with microfluidic systems to create artificial circulatory systems, delivering fuel and removing waste products, enabling sustained bioluminescent activity. Think of a self-sustaining artificial organism, powered by chemical reactions within its microfluidic network, its bioluminescent skin pulsating with the rhythm of its internal processes. Such a system would blur the lines between machine and organism.

Moving beyond individual organisms, we can envision complex ecosystems of interacting artificial life forms. These ecosystems could be designed to perform specific tasks, like breaking down pollutants or producing biofuels. The bioluminescent signals of individual organisms could be coordinated to facilitate communication and cooperation within the ecosystem, creating a mesmerizing symphony of light. Imagine a closed-loop system where artificial organisms, powered by sunlight, convert waste materials into biofuels, signaling their progress through synchronized flashes of bioluminescence.

The development of advanced artificial organisms requires close collaboration between multiple disciplines. Computer scientists are developing sophisticated algorithms to control the behavior of these organisms. Materials scientists are creating novel materials with biocompatible properties. Synthetic biologists are engineering bioluminescent pathways that can be integrated into artificial systems. This interdisciplinary approach is essential to realize the full potential of artificial life. Consider a team of researchers, each with their specialized expertise, working together to create an artificial organism capable of mimicking the complex bioluminescent patterns of a deep-sea jellyfish.

The spectral properties of the emitted light are crucial for various applications. For example, in underwater communication, the wavelength of the bioluminescent signal must be carefully chosen to minimize absorption by the surrounding water. In bio-inspired robotics, the spectral output can be used to camouflage the robot or to attract prey. The ability to precisely control the spectrum of artificial bioluminescence opens up a wide range of possibilities. Imagine an artificial organism designed to interact with a specific species of deep-sea fish, its bioluminescent signal perfectly matching the fish's visual sensitivity, enabling targeted communication and interaction.

Furthermore, the temporal dynamics of bioluminescence are just as important as the spectral characteristics. The frequency and duration of light pulses can convey information, enabling communication between artificial organisms and even with natural organisms. Researchers are exploring ways to create artificial organisms that can modulate their bioluminescent signals in complex ways, mimicking the intricate flashing patterns observed in fireflies or the rhythmic pulsations of jellyfish. Consider a swarm of artificial organisms communicating through carefully timed flashes of bioluminescence, coordinating their movements and actions in a mesmerizing display of synchronized behavior.

The ethical implications of creating advanced artificial organisms must be carefully considered. As these entities become more sophisticated and lifelike, questions about their moral status and potential impact on the environment become increasingly relevant. It is crucial to establish ethical guidelines for the development and deployment of artificial life, ensuring that these technologies are used responsibly and for the benefit of humanity and the planet. This requires ongoing dialogue between scientists, ethicists, policymakers, and the public. Imagine a future where artificial organisms play a significant role in our society, raising complex ethical questions about their rights and responsibilities, prompting us to redefine our relationship with technology and the natural world.

Finally, the creation of advanced artificial organisms presents an unprecedented opportunity to gain a deeper understanding of the origins and nature of life itself. By building artificial systems that exhibit lifelike behaviors, we can explore the fundamental principles that govern living systems and potentially shed light on the very essence of what it means to be alive. This pursuit of knowledge has the potential to transform our understanding of biology, evolution, and our place in the universe. It compels us to ask profound questions about the nature of consciousness, intelligence, and the future of life itself.

7.2 Synthetic Bioluminescence

The quest to create synthetic bioluminescence represents a significant convergence of biology, chemistry, and engineering. It goes beyond simply mimicking naturally occurring bioluminescence; it aims to engineer novel light-emitting systems with tailored properties. This involves not just replicating existing luciferin-luciferase pathways, but also designing entirely new ones. Imagine bioluminescent plants illuminating our streets, or self-powered bioluminescent displays integrated into clothing. This chapter delves into the exciting possibilities and challenges of this burgeoning field.

One promising approach in synthetic bioluminescence involves modifying existing biological systems. Scientists are exploring ways to introduce bioluminescent genes into non-luminous organisms, effectively creating glowing plants or animals. This has been achieved with some success in plants like tobacco and Arabidopsis, offering a glimpse into the potential for bioluminescent crops and even trees. Challenges remain, however, in optimizing the intensity and duration of the emitted light. Furthermore, scaling up production for practical applications requires careful consideration of energy efficiency and environmental impact.

Beyond modifying existing organisms, synthetic biology allows for the creation of entirely new bioluminescent systems from the ground up. This involves designing novel enzymes and substrates that can react to produce light. By tweaking the molecular structures of these components, researchers can fine-tune the color and intensity of the emitted light. This opens up exciting possibilities for creating custom-designed bioluminescent materials with specific applications in mind, from medical imaging to biosensors. The complexities of biological systems, however, often make this a challenging endeavor.

Nanotechnology plays a crucial role in synthetic bioluminescence. Nanoparticles can be used to encapsulate bioluminescent molecules, protecting them from degradation and enhancing their light output. These nanoparticles can also be functionalized with targeting moieties, enabling them to deliver bioluminescent signals to specific locations within the body. Imagine, for instance, using bioluminescent nanoparticles to track the spread of cancer cells or monitor drug delivery in real-time. The precise control over nanoparticle size and surface chemistry is key to realizing the full potential of this approach.

Another emerging area of research focuses on harnessing the power of bacterial bioluminescence. Certain bacteria naturally produce light through a complex enzymatic pathway. Scientists are investigating ways to engineer these bacterial systems to produce light on demand and in response to specific stimuli. This could lead to the development of self-powered biosensors that detect environmental pollutants or diagnose diseases. The challenge lies in understanding the intricate regulatory mechanisms that control bacterial bioluminescence and adapting them for practical applications.

Fungal bioluminescence, while less studied than bacterial or marine systems, also holds promise for synthetic biology. Certain fungi emit a continuous, soft glow, unlike the flashes of light produced by many other bioluminescent organisms. This continuous luminescence makes them attractive candidates for creating self-illuminating materials. Researchers are working to understand the genetic basis of fungal bioluminescence and develop methods to control its intensity and color. Unlocking the secrets of fungal bioluminescence could pave the way for sustainable and energy-efficient lighting solutions.

Looking ahead, synthetic bioluminescence has the potential to revolutionize various fields. In medicine, it could provide new tools for non-invasive imaging and diagnostics. In environmental science, bioluminescent sensors could monitor pollution levels and track the movement of pollutants in real-time. In materials science, bioluminescent materials could create self-powered displays and lighting systems. The possibilities are vast and limited only by our imagination and our ability to harness the power of biology and chemistry. The key to unlocking this potential lies in continued research and development, pushing the boundaries of what is possible with synthetic bioluminescence.

The ethical implications of synthetic bioluminescence must also be considered. As with any powerful technology, there is the potential for misuse. Guidelines and regulations are necessary to ensure responsible development and application of this technology. Open discussions and collaborations between scientists, ethicists, and policymakers are crucial to navigate the ethical landscape and maximize the benefits of synthetic bioluminescence for society. These discussions should address issues such as environmental impact, potential health risks, and the responsible use of genetically modified organisms.

Synthetic bioluminescence is not merely a scientific curiosity; it is a powerful tool with the potential to reshape our world. By understanding the intricacies of natural bioluminescence and applying the principles of synthetic biology, we can create novel light-emitting systems with tailored properties. This journey of discovery promises to illuminate not just our understanding of biology, but also our path towards a brighter and more sustainable future. The challenges are significant, but the potential rewards are immense, making synthetic bioluminescence a field ripe with opportunity and innovation. The ongoing research in this field holds the key to unlocking a future where light is not just a source of illumination, but also a powerful tool for advancing science, medicine, and technology.

7.3 Interspecies Communication

The potential for interspecies communication using bioluminescence represents a frontier of scientific exploration. Imagine a future where engineered organisms, equipped with synthetic bioluminescent systems, can engage in dialogue with naturally bioluminescent species. This chapter delves into the nascent field of bioluminescent communication, exploring the possibilities, challenges, and potential implications of bridging the communication gap between humans and the natural world. This involves understanding the nuances of bioluminescent signaling in different species, deciphering the "language" of light pulses, and developing technologies to generate and interpret these signals.

Current research focuses on identifying the specific light patterns and their associated meanings in various bioluminescent organisms. Fireflies, for example, use precisely timed flashes for courtship, while deep-sea creatures employ complex light displays for camouflage, prey attraction, and communication. Deciphering these intricate codes is crucial for establishing any form of meaningful dialogue. By analyzing the spectral properties, temporal patterns, and spatial distribution of bioluminescent signals, researchers are beginning to unravel the syntax of this luminous language. This involves developing sophisticated algorithms and machine learning models to recognize and classify different signal types, ultimately creating a "dictionary" of bioluminescent communication.

Beyond mere observation, the next step involves developing technologies capable of generating artificial bioluminescent signals that mimic natural patterns. This requires engineering synthetic bioluminescent systems with precise control over light intensity, color, and flash frequency. Researchers are exploring various approaches, including genetically modifying existing bioluminescent organisms and designing entirely new synthetic systems based on nanotechnology and materials science. The goal is to create a versatile toolkit for generating customized light signals that can be "understood" by different species. Imagine an artificial organism emitting a specific sequence of light pulses that attracts a particular species of deep-sea squid, allowing researchers to observe its behavior up close or even collect samples without invasive capture methods. Such targeted communication could revolutionize our understanding of elusive deep-sea ecosystems.

The applications of interspecies communication extend beyond scientific research. In environmental monitoring, engineered bioluminescent bacteria could be deployed to detect pollutants in water or soil, emitting specific light signals in response to different contaminants. This would provide real-time, highly sensitive environmental monitoring without the need for complex laboratory analysis. In agriculture, bioluminescent plants could signal nutrient deficiencies or pest infestations, enabling targeted interventions and reducing the need for widespread pesticide application.

However, the potential of interspecies communication also raises important ethical considerations. The ability to manipulate bioluminescent signals could have unintended consequences for ecosystems. For instance, introducing artificial signals into a natural environment could disrupt established communication patterns, affecting mating behavior, predator-prey dynamics, and overall ecosystem stability. Furthermore, the development of sophisticated bioluminescent communication technologies raises concerns about potential misuse, such as manipulating animal behavior for human gain. Careful consideration and ethical guidelines are crucial to ensure responsible development and deployment of these powerful tools.

One fascinating area of exploration is the possibility of creating symbiotic relationships between artificial organisms and naturally bioluminescent species. Imagine an artificial organism integrated into a coral reef ecosystem, communicating with the resident bioluminescent organisms to monitor and maintain the health of the reef. The artificial organism could detect signs of stress, such as changes in water temperature or acidity, and communicate this information to the corals, triggering adaptive responses or alerting human researchers to the problem. Such symbiotic partnerships could be crucial for preserving fragile ecosystems in the face of climate change and other environmental threats.

Research in this field is not without its challenges. Bioluminescent systems are incredibly complex, and replicating their intricate functionality in artificial organisms is a formidable task. Furthermore, the vast diversity of bioluminescent organisms, each with its own unique communication code, presents a significant hurdle to establishing universal communication protocols. Despite these challenges, the potential rewards of interspecies communication are immense, offering a glimpse into a future where humans can engage in meaningful dialogue with the natural world. This new form of communication could revolutionize our understanding of life on Earth, enabling us to learn from and coexist with the diverse array of bioluminescent organisms that illuminate our planet. The journey towards achieving this ambitious goal is still in its early stages, but the path forward is illuminated by the promise of a deeper connection with the living world around us. The quest to decipher the language of light promises to unlock a wealth of knowledge and open up exciting possibilities for interaction with the natural world. As we continue to explore the intricacies of bioluminescence and develop increasingly sophisticated technologies for generating and interpreting light signals, we move closer to realizing the dream of true interspecies communication. This journey will undoubtedly require interdisciplinary collaboration between biologists, chemists, engineers, and ethicists to ensure responsible and beneficial outcomes.

7.4 Ethical Considerations

The convergence of artificial organisms and bioluminescence presents a novel frontier in scientific exploration, raising profound ethical questions that demand careful consideration. As we delve deeper into mimicking and manipulating natural processes, particularly those as intricate and delicate as bioluminescence, we must grapple with the potential consequences of our actions. This exploration isn't simply about technological advancement; it necessitates a thorough examination of our responsibilities and the potential impact on the environment, existing organisms, and society.

One immediate concern arises from the potential disruption of ecosystems. Introducing artificial organisms, especially those designed for environmental monitoring or interaction, could have unforeseen consequences on existing biological communities. Competition for resources, unintended predation, or the introduction of novel light patterns could disrupt established ecological balances. Consider the impact on nocturnal animals that rely on natural bioluminescence for communication, mating, or predator avoidance. The introduction of artificial light sources, even those intended to mimic natural bioluminescence, could interfere with these delicate processes. Therefore, rigorous ecological risk assessments are essential before deploying artificial bioluminescent organisms in natural environments. We must prioritize the preservation of existing biodiversity alongside our pursuit of scientific discovery.

Furthermore, the potential for misuse of this technology warrants close scrutiny. The ability to create artificial organisms with controlled bioluminescent properties raises concerns about potential applications in surveillance, camouflage, or even manipulation of natural behaviors. Imagine the implications of deploying artificial organisms capable of mimicking the distress signals of prey animals to lure predators, or conversely, using them to create distracting light displays to disorient prey. Such applications could have significant and potentially detrimental impacts on natural ecosystems and food chains. Establishing clear ethical guidelines and regulations for the development and deployment of these technologies is crucial to preventing their misuse.

The very act of replicating and potentially modifying natural bioluminescence raises questions about our relationship with the natural world.

Bioluminescence is a product of millions of years of evolution, representing a complex and delicate interplay of biological processes. Our ability to recreate this phenomenon artificially raises questions about our responsibility to respect and preserve the natural world. Are we justified in manipulating these natural processes for our own purposes? What are the potential long-term consequences of blurring the lines between natural and artificial life? These are complex philosophical questions that deserve thoughtful consideration.

Beyond ecological concerns, ethical considerations extend to the societal implications of these advancements. The potential for commercial exploitation of artificial bioluminescence raises concerns about intellectual property rights, equitable access to the benefits of this technology, and the potential for widening existing social inequalities. Who owns the rights to these synthetically created organisms and their unique light emissions? How do we ensure that the benefits of this technology, such as improved environmental monitoring or medical applications, are accessible to all, not just a privileged few? Open dialogue and careful consideration of these societal implications are crucial to shaping a future where this technology benefits humanity as a whole.

Finally, the ethical considerations surrounding artificial bioluminescence necessitate a nuanced approach to risk assessment and regulation. Traditional risk assessment frameworks may not adequately address the unique challenges posed by these emerging technologies. The complex interactions between artificial organisms, natural ecosystems, and societal values demand a more holistic approach. This requires ongoing dialogue between scientists, ethicists, policymakers, and the public to ensure that the development and deployment of these technologies are guided by ethical principles and contribute to a sustainable and equitable future. The ethical dimensions of this field are not merely an afterthought but an integral part of the scientific endeavor, demanding our continuous attention and reflection. Only through careful consideration of these ethical considerations can we navigate the complexities of this exciting field and harness its potential for the benefit of both humanity and the natural world.

7.5 Open Questions

The convergence of artificial organisms and bioluminescence generates a cascade of inquiries that reach beyond the current boundaries of our understanding. While we have made strides in simulating bioluminescent processes and analyzing spectral overlaps, the intricate interplay between these two domains remains largely uncharted. One fundamental question revolves around the potential for genuine communication between artificial organisms and naturally bioluminescent life forms. Can we engineer artificial light emissions that elicit specific responses from bioluminescent bacteria, fungi, or marine organisms? Could this lead to novel forms of interspecies communication, enabling us to interact with and potentially influence the behavior of these organisms in their natural environments? This raises further questions about the ecological implications of introducing artificial bioluminescent entities into existing ecosystems.

The long-term effects of such interactions remain unclear. How might the introduction of artificial bioluminescent signals impact the established communication networks within these ecosystems? Could it disrupt existing symbiotic relationships or create unintended ecological imbalances? Detailed ecological studies and carefully controlled experiments are essential to address these concerns before widespread deployment of artificial bioluminescent organisms. Beyond ecological considerations, the ethical implications of engineering artificial organisms capable of interacting with naturally bioluminescent life warrant careful scrutiny. Do we have the right to introduce synthetic light signals into environments where they may influence the behavior and survival of existing species? What are the potential long-term consequences of manipulating these delicate ecosystems, and who bears the responsibil-

ity for unforeseen outcomes?
Moving beyond ecological and ethical considerations, we encounter a series of fascinating scientific questions that remain unanswered. The mechanisms underlying spectral entrainment and synchronization in bioluminescent systems are not fully understood. What are the specific molecular and biophysical processes that allow for the coordinated flashing patterns observed in fireflies or the synchronized waves of light produced by dinoflagellates? Can we replicate these intricate mechanisms in artificial organisms, potentially leading to the creation of self-organizing, bioluminescent swarms capable of performing complex tasks? Investigating these fundamental processes requires interdisciplinary collaborations between biologists, chemists, physicists, and engineers. Another area ripe for exploration lies in the potential for artificial bioluminescence to serve as a novel tool for biomedical applications. Could we engineer bioluminescent artificial organisms capable of targeting specific cells or tissues within the human body, providing real-time imaging or delivering therapeutic payloads?
The possibility of using bioluminescence for in vivo diagnostics and targeted drug delivery represents a promising avenue for future research. The development of advanced artificial organisms with sophisticated bioluminescent capabilities raises further questions about the potential for creating entirely new forms of life. Can we engineer artificial organisms that exhibit not only bioluminescence but also other essential biological functions, such as self-replication and adaptation? The ethical considerations surrounding the creation of synthetic life are profound and require careful consideration before embarking on such ambitious endeavors. Furthermore, the exploration of spectral synchronicity between artificial organisms and bioluminescence opens up possibilities for harnessing this phenomenon for technological advancements. Can we utilize the principles of spectral entrainment to develop novel communication technologies, secure encryption methods, or even new forms of energy transfer? The potential applications of this emerging field are

vast and remain largely unexplored.

Finally, the study of spectral synchronicity between artificial organisms and bioluminescence compels us to reconsider our understanding of life itself. What are the fundamental principles that govern the emergence of complex behavior in both natural and artificial systems? Can the study of artificial life illuminate the origins and evolution of life on Earth? These fundamental questions challenge us to push the boundaries of our scientific understanding and explore the profound implications of creating artificial life forms that mimic and interact with the natural world. Exploring these open questions requires not only scientific ingenuity but also a deep awareness of the ethical and societal implications of this rapidly evolving field. As we delve deeper into the interplay between artificial organisms and bioluminescence, we must remain mindful of the potential consequences of our actions and strive to harness this knowledge for the benefit of both humanity and the planet. The answers we seek may not only revolutionize our understanding of life but also shape the future of our relationship with the natural world.

www.ingramcontent.com/pod-product-compliance
Lightning Source LLC
LaVergne TN
LVHW022014200926
848481LV00007B/2392
* 9 7 9 8 3 4 8 3 9 2 3 7 6 *